HEART of the HIVE

Heart of the HIVE

Inside the Mind of the Honey Bee and the Incredible Life Force of the Colony

HILARY KEARNEY

Photography by **ERIC TOURNERET**

Storey Publishing

The mission of Storey Publishing is to serve our customers by publishing practical information that encourages personal independence in harmony with the environment.

EDITED BY Deborah Burns and Hannah Fries
ART DIRECTION AND BOOK DESIGN BY Bredna Lago
TEXT PRODUCTION BY Jennifer Jepson Smith
COVER AND INTERIOR PHOTOGRAPHY BY © Eric Tourneret, except for page 181 by © Tosca Radigonda
ILLUSTRATIONS BY © Hilary Kearney

Storey Publishing
210 MASS MoCA Way
North Adams, MA 01247
storey.com

Storey Publishing is an imprint of Workman Publishing, a division of Hachette Book Group, Inc., 1290 Avenue of the Americas, New York, NY 10104. The Storey Publishing name and logo are registered trademarks of Hachette Book Group, Inc.

ISBNs: 978-1-63586-483-0 (hardcover); 978-1-63586-484-7 (ebook)

Printed in China through Asia Pacific Offset
10 9 8 7 6 5 4 3 2 1

Library of Congress Cataloging-in-Publication Data on file

Cover image: A worker bee passes nectar to her hivemate.

For my three-year-old son, Kit.
May you find a passion on par with the one I have for bees.
Of course, I'll love you no matter what, but ideally,
your passion will also be bees. I need help carrying supers.

Contents

A worker bee fills the cells of some newly constructed honeycomb.

Introduction

In stories, hidden worlds are found in wardrobes and down rabbit holes, but I found mine in a beehive. It's a wonderland I can enter with the zip of my bee suit. A golden realm, vibrating and fragrant, full of treasures, surprises, and mysteries. Its society is enchanting and complex, completely different from our own, but when I visit, I feel at home.

In my quest to know honey bees, I've discovered that magic is real. It's all around us, in the details of nature. I once believed that science and magic were opposites, that an understanding of life's mysteries somehow lessened the wonder, but now I know that wonder expands with knowledge. Every discovery leads to more questions and takes us deeper into enchantment.

If I look closely into any subject, I can find a reason to marvel. After all, the prismatic beauty of honey bee wings can also be admired in the wings of flies. Even so, there are few creatures as fascinating and vital as the honey bee. She is astoundingly complex, an individual, but also part of a larger animal—her colony. Independently, she can understand numbers, predict the weather, and engage in abstract thought. She flies miles from her hive each day and still finds her way home. Together, she and some 50 thousand sisters are even more impressive. Organized by a complex job system, they spend their lives working for the good of the colony and achieve a kind of immortality. Yet the benefit of their efforts extends far beyond the hive. They are the world's most popular pollinators, contributing not only to agriculture but also to entire ecosystems. It's no wonder those who work closely with honey bees revere them.

Our connection to honey bees is ancient and deep, a resounding hum that has continued for at least nine thousand years and a kinship that has influenced agriculture, medicine, architecture, religions, and more. Yet few people know the details of a honey bee's life. As a beekeeper, I am compelled to teach others about the importance of honey bees, but I want to do more. I want to take people tumbling down the rabbit hole.

Why Honey Bees?

Everyone, not just the beekeeper, should know honey bees. They are pivotal to life as we know it and shape the landscapes we share. Because they are so deeply connected to the ecosystem, an understanding of them can only expand our awareness of all living things and remind us of our responsibilities to this world. Thanks to their adaptability and their partnership with humans, honey bees now thrive on every continent except Antarctica. Thousands of them emerge from their hives every day, flying out into their surroundings to forage on the nectar and pollen they need to survive. In doing so they become harbingers of life, dusting the flowers they visit with pollen so that each one may make seeds. If we were to trace the path of every departing honey bee, we would see an enormous, crisscrossing mass that extends miles in each direction. This network of flying foragers is responsible for pollinating an astounding number of plants. Honey bees may be the single most prolific pollinators on Earth.

Often recognized for their contributions to agriculture, honey bees are responsible for pollinating more than 130 varieties of fruits, nuts, and vegetables in the US alone. Yet they are also the most frequent floral visitors in natural habitats worldwide. A recent survey of wild plant pollination found that honey bees were the exclusive pollinators of 5 percent of the observed plants—a huge amount for just one species. But they cannot do it alone. An estimated 87 percent of flowering plants on Earth rely on animals for pollination. Although honey bees are clearly the heavyweights of the pollinator category, they did not visit every plant type. In fact, they ignored 49 percent of the observed plants! This highlights the critical role native pollinators play in their ecosystems.

Bee Diversity

Most bees are not like honey bees. They don't live in colonies or make honey—they may not even have stripes! Some are covered in snowy, gray fuzz, while others shine like emeralds. The largest bee, *Megachile pluto*, is close to the size of a walnut, while the smallest bee, *Perdita minima*, is roughly the size of a chia seed.

SOLITARY BEES

The typical bees' nest is not a busy tree hollow packed with bee-covered honeycomb but a modest hole in the ground. Approximately 75 percent of bees are solitary, and the majority nest in soil. Some live in hollow plant stems, while others have adapted to their human-dominated environments by nesting in patio furniture or even wind chimes! The mother bee simply needs a protected tunnel to raise her young. Inside she lays each egg on a pillow of pollen large enough to feed it through the larval stage. In most species, the mother bee never meets her offspring. Her life is as brief as the flowers she pollinates, but her young will emerge the following year, like waiting seeds, and begin the cycle again.

SOCIAL BEES

Few bees belong to this exclusive social club. Only honey bees, bumble bees, and stingless bees live in colonies with workers and a queen (or queens), but they have fascinating differences.

A bumble bee colony is founded by a solitary queen and lives only for a year. After hibernation, the queen builds a soft nest and lays her first batch of eggs. The eggs become worker bees that help her to raise more eggs and build comb. In comparison to honey bee comb, it looks bubbly and disorganized, like someone glued a bunch of Kix cereal puffs together. In late summer the colony raises new queens in hopes that they will overwinter and establish new nests the following spring.

Stingless bees are only found in tropical regions. These tiny bees can establish large nests with populations up to 40 thousand. Like those of honey bees, stingless bee colonies are perennial and have queens, workers, and drones. They are sometimes kept by humans, and although they produce much less honey, they are cherished as pets.

It is my hope that this book will not only unveil the hidden world of honey bees but also awaken you to the possibility of the thousands of other amazing species deserving of our attention. Perhaps diving into the world of honey bees will ignite a curiosity about other bee species and pollinators, but I hope it also challenges you to look at all creatures in a different light. Let the honey bee be your gateway bug. I believe that every wasp, rat, and pigeon is worthy of study, respect, and celebration.

Melipona beecheii stingless bees were once kept and revered by the ancient Maya. The tradition continues today in the Yucatán peninsula of Mexico.

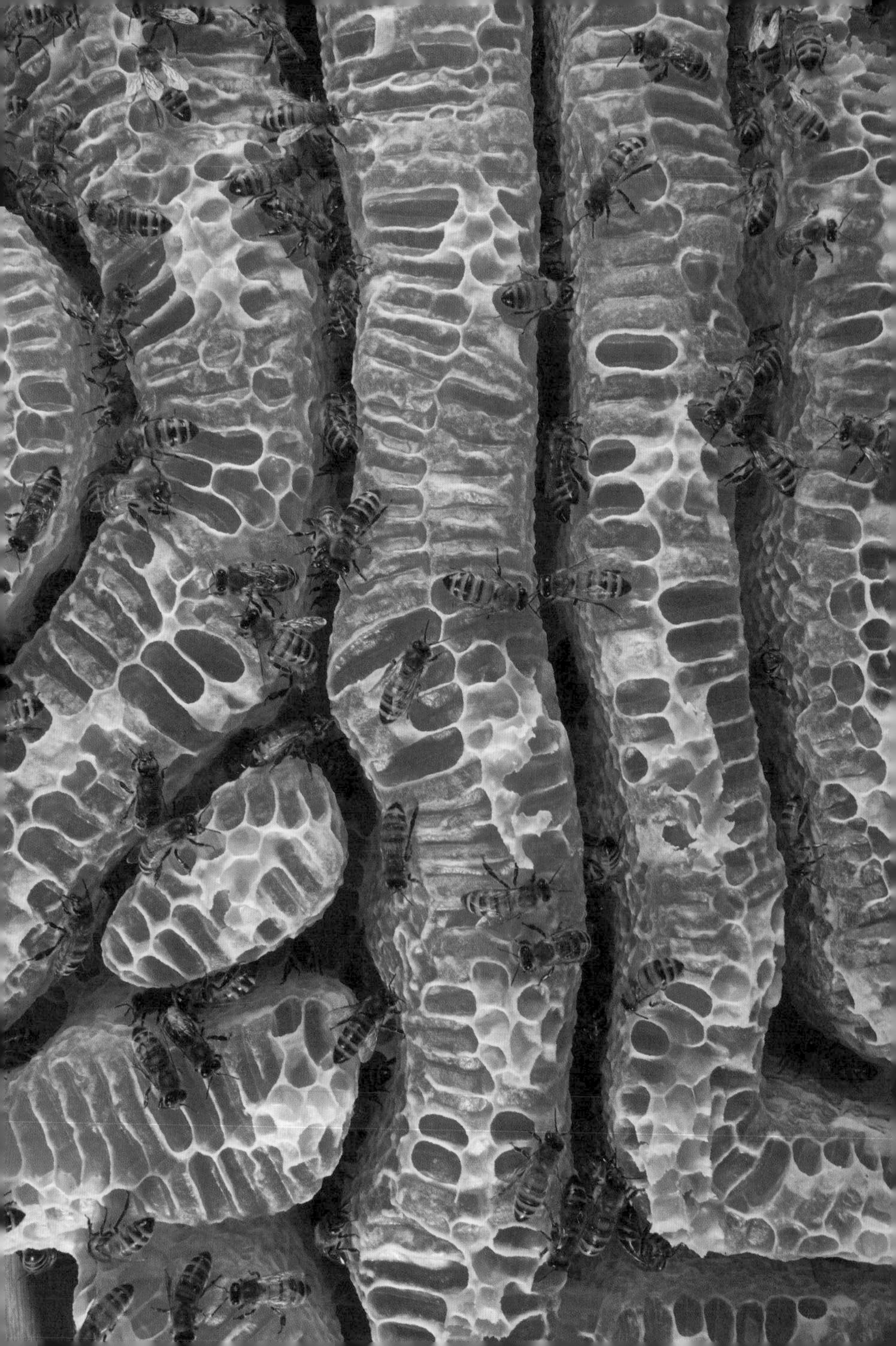

CHAPTER 1

Life as a Super-organism

How the cooperative nature of the beehive makes honey bees unique

There are more species of bees on Earth than there are mammals and birds put together. Yet even among more than 20 thousand species of bees, honey bees are unique. With many living as one, the beehive is a golden monument to social cooperation. Their intricate way of life is steeped in mystery and feels almost beyond comprehension. This is where the magic lies.

When left to their own devices, wild honey bee colonies often build their honeycomb in beautiful, maze-like patterns.

The Infinite Animal

The honey bee is more than herself. She is part of a wondrous other—the superorganism. Unlike most creatures, she is not driven by an individual need to survive and reproduce. In fact, she can't live on her own. She is part of a body of fellow worker bees who, with a queen and a cohort of brothers, dedicate themselves to the survival of the colony. She will live only a few weeks, but her colony can endure indefinitely. She is an infinite animal that constantly adapts and changes with every new generation of bees. Even so, she maintains a oneness, a kind of central life force.

Beekeepers will tell you that each colony acts almost like an individual. Each has unique traits and temperaments that persist long after the founding members are gone. Still, we struggle to understand the oneness of honey bee colonies. We often think of a stinging honey bee as altruistic. She sacrifices her life to defend the colony, but perhaps she sees no difference between herself and the whole.

On the other hand, individual bees may still have distinct personality traits. Researchers discovered that some bees are more adventurous than others. These brave bees prefer to work as scouts and will seek out new food sources and nesting sites for their colonies. Bees with this penchant for new experiences have distinct brain patterns that are also found in humans with thrill-seeking personalities. It seems that the honey bee is a transcendent being. She is somehow simultaneously her unique self and part of a collective, enduring soul. Her way of life is so alien to us, it's no wonder Hollywood borrows the superorganism for so many of their sci-fi flicks.

Comb: The Body

Comb is a curious combination of architecture and biology. It's a structural nest used to raise young and store food, but it's also the body of the superorganism. If the honey bee colony is one animal, then its comb serves as its skeleton, internal organs, and immune system. The bees cannot function without it.

The first combs built in each hive are used primarily to raise new bees. These nursery combs are at the center of the hive and must be kept warm for the new bees to develop properly. A dense crowd of adult bees climbs up and down the nursery combs like ladders as they tend to the young and maintain the temperature. This part of the hive is more than a nursery: It is the womb of the superorganism. Each cell is a cradle for an infant larva, but its thin walls also work like an amniotic sac. Sealed inside the cell, the larva is protected from harmful microbes as she grows. Once she's fully developed, she births herself by breaking through the cell cap—and emerges into her hive.

The combs that surround the brood make up the pantry of the beehive. The bees use the same methods as human homesteaders to preserve their harvests. Each cell is like a mason jar full of precious food, either pollen or honey. Pollen is fermented in the cells using bacteria and yeast to keep it from spoiling, while flower nectar is preserved through dehydration, a process that turns it into honey.

In addition to functioning like farmhouse kitchens, these combs also act like body fat. Not only do they store and release energy (i.e., food) for the superorganism, they also insulate the colony. Like blubber, large swaths of honeycomb make up the outer layers of the hive and help to keep the inner brood combs warm by preventing heat loss.

The Workers: The Blood

Worker bees are the lifeblood of the colony. This all-female workforce represents 75 to 100 percent of the bees in the hive, depending on the time of year. Most live only six weeks, but together they perform the various tasks that keep the superorganism up and running. They are organized by a flexible job system that constantly adapts to the colony's changing needs.

Donating Blood

When a colony is failing in the apiary, I often advise new beekeepers to increase the number of adult bees. Beekeepers with more than one colony can achieve this by taking capped brood from a strong colony and donating it to the weak one. The brood at this stage no longer needs to be fed, so it is not very taxing on the already struggling colony. The nurse bees need only to keep the brood warm long enough for the near-ready bees to emerge. Since they emerge inside the hive, the new bees are accepted and quickly utilized even though they are not genetically related to the rest of the bees. In a typical Langstroth hive, a frame of capped brood contains about six thousand developing bees, and this burst of new workers can act like a lifesaving blood transfusion to the colony.

In some ways worker bees are like the actual blood of the superorganism. They circulate continuously through the combs, bringing in fresh oxygen and fanning out carbon dioxide. They disperse nutrients, remove waste, and fight infections. The colony also requires a certain volume of workers to function. New workers are continuously raised to replace older ones, but if the worker population becomes too low, the colony goes into a kind of organ failure and will not survive. Individual worker bees may be expendable, but as a group they are vital.

The Queen: Life Force

The queen bee is the main reproductive force in the colony and the only female who can mate. She can lay her own body weight in eggs every day, 1,500 or more! In this way, she is the ovaries of the superorganism, yet her role is more vital than that. Since there is typically only one queen in a colony, she alone shoulders the responsibility of keeping the colony's total population up. She also acts as the marrow of the superorganism, a central life force that generates new bees and motivates them to perform their duties. Without her, the colony is doomed to fail.

The queen's lifespan is much longer than that of the other bees in the hive. She typically lives two to three years, but some queens have lived to be eight years old. As she ages, her fertility and pheromones wane, prompting the worker bees to raise a replacement queen. They do so by constructing a special chamber, a queen cell, which they fill with a nutrient-rich food called royal jelly. This milklike substance is generated by the worker bees. A larva that is fed royal jelly exclusively will develop into a new queen bee. (More on page 15.)

The Drones: The Sperm

If the queen is the ovaries of the superorganism, then the drones are the sperm. Indeed, they have many traits in common with actual sperm. Their singular goal is to pass on their colony's genes, and they must leave the body of the colony to do so. During mating season, a colony expels thousands of drones from the hive in hopes they will succeed in mating with another colony's queen. Yet a colony only produces drone comb when it reaches a population of four thousand or more worker bees. In this way, we can think of the colony as an organism that must wait to reach sexual maturity before it attempts to mate with other colonies via drones.

Honey: Sustenance

Made from the nectar of millions of flowers, honey, some say, is the culmination of sunshine itself. Yet every drop begins in the body of a bee and is passed from bee to bee before it can be called honey. The flowers provide the ingredients, but it is the bees who make the honey. Honey

A newly laid honey bee egg stands on end in its cell.

is an important part of the colony's diet. It is the bees' main source of carbohydrates, and it provides them with the energy they need to fly, make beeswax, and perform daily maintenance activities for the hive. Honey making is a truly collaborative process. It's estimated that each individual bee makes about seven drops of honey in her lifetime. This amount would fill only two and a half cells, but the colony as a whole can make hundreds of pounds of honey per year.

As soon as a bee extends her tongue into a flower and takes up the sweet nectar, the nectar is changed. She fills a special organ inside her body called the honey stomach. This expandable organ functions like an internal grocery bag that she uses to carry liquids back to the beehive. While the nectar is inside her, it is mixed with enzymes that transform it. A key enzyme called invertase converts sucrose—a complex sugar found in both nectar and table sugar—to simple sugars: glucose and fructose. These simple sugars are easier for humans to digest and make the honey taste sweeter than table sugar.

When a bee returns to the hive filled with nectar, she passes it to another bee. She regurgitates the nectar and holds the tiny droplet with her mandibles. The receiving bee extends her tongue and begins filling her own honey stomach. In this way, the nectar travels from bee to bee. Each time it is passed, the nectar is exposed to air and loses some of its water content. It also gains additional enzymes and may be a way for bees to share healthy gut microbes with each other.

Each nectar load may pass through hundreds of bees before it reaches a honeycomb cell, becoming more and more concentrated as it does. The bees then need to reduce it from its original state of about 80 percent water to just 18 percent. So they spread the nectar on the cell walls to increase its surface area and then fan their wings to speed the process. On summer evenings, beekeepers can watch fanning bees at their hive's entrance as they pull out the humid air to dry the nectar inside. The intoxicating honey aroma is so strong, it can be smelled from several feet away.

Once the honey is at the correct moisture level, the bees move it from the drafty combs near the entrance to storage combs above the brood nest. There, the honey is sealed into the cells with a thin layer of beeswax. The honey cannot spoil, and the bees can eat it as needed through the winter.

Bee Bread: Nutrition

Pollen is gathered with style. An accomplished pollen forager may carry so much colorful pollen on her back legs that she starts to resemble an MC Hammer backup dancer. Yet somehow she is still able to fly home.

Perfectly Preserved

Despite the astounding longevity of Twinkies, honey is the only food on Earth that doesn't spoil. Its low water content, high sugar content, and acidic pH make it impossible for bacteria and other harmful microbes to survive in it. It also contains other antimicrobial components, such as hydrogen peroxide. Honey is a potent antimicrobial substance—it's even effective against multidrug-resistant bacteria.

Honey has been used for thousands of years in medicine and, to this day, is used to treat burns, wounds, and inflammation. Its low pH enhances tissue repair and reduces scarring. Yet no single element is responsible for its miraculous healing effects. Studies have shown that, like the bees themselves, the different components in the honey work synergistically.

Once inside her hive, she unloads her pollen pants directly into a cell near the brood nest. The pollen is packed down and layered with saliva and nectar between each load.

From the side, a filled pollen cell looks like the colorful layers of sediment found in desert sandstone, each thin band from a different flower type. Here the raw pollen is fermented into a protein-rich food that beekeepers call bee bread. This nutritious, doughlike food contains important amino acids, minerals, and fats and is critical to the development of young bees.

The fermentation process begins with *Lactobacillus* bacteria that metabolize the fructose in the pollen/nectar mixture and produce lactic acid. The lactic acid reduces the pH, which creates the perfect environment for yeast to thrive. The bees rely on the same genus of yeast we use to make our own foods, including sourdough bread and beer. The fermented bee bread lasts much longer than raw pollen because the beneficial yeast prevents other harmful microbes from growing and spoiling the food.

Larval Jelly: Bee Milk

Honey bees have an unexpectedly mammalian trait—they produce a kind of milk to feed their young. Worker bees charged with caring for developing bees excrete a nutritious, milky-white jelly from special glands in their heads and feed it to hungry larvae (see page 28). This larval jelly is most famously known as royal jelly because of its role in the queen bee diet, but larval jelly is fed to all three honey bee castes: workers, drones, and queens. However, each caste receives its own special formula and in differing amounts, so there is not only royal jelly but also worker jelly and drone jelly. Collectively, these are known as brood food or brood jelly.

Propolis: Skin

Propolis is a multipurpose substance that the bees make from plant resins and beeswax. This sticky mixture is used as a building material, but it also is closely linked with colony health. To make propolis, a forager bee collects resins from nearby plants. These plants produce various saps and oils to protect themselves from disease. The forager scrapes the resin off the plant and packs it onto her legs, like she would pollen, to carry it back to her hive. There, nestmates help remove the sticky substance from her legs and begin to use it in the hive. It is sometimes mixed with beeswax, but its chemical composition isn't changed much.

Propolis is the reason that beekeepers need what is essentially a mini crowbar to break into their hives. The bees glue all the hive boxes together with propolis, and the bond can be incredibly strong. Once the hive is open, beekeepers can easily see the thick line of propolis that held the hive components together. Sometimes it's so thick that it stands up like pulled taffy. Propolis may have a brown, red, or even green coloration; it depends on what type of plants the bees collected resin from.

Propolis has a more important function, however, than annoying beekeepers; it is also antimicrobial. The bees coat the whole inside of their hive cavity with a layer of it. In hives constructed by humans, smooth milled wood appears to discourage the bees from applying very much propolis. Instead of the thick lacquered layer found in a natural hive, there is only a thin stain. Still, propolis acts like the superorganism's skin. It is a barrier that protects the collective body from harmful

When pollen is stored in the comb, its striking colors are on display. This rainbow of pollen represents the diversity of flowers blooming near the hive.

microbes. Like skin, which contains immune cells that fight invading organisms, propolis can kill bacteria, fungi, and even viruses. The bees also use propolis as a kind of immune response and will employ it to entomb unwelcome guests who try to enter their hive, thus preventing the growth and spread of harmful bacteria. For example, if an unlucky mouse is stung to death inside the hive, the bees will use propolis to mummify its body.

Lifecycle of the Individual

Every bee begins as an oval, pearlescent egg. The queen bee uses her elongated abdomen to deposit each egg at the bottom of its own cell. If the egg is destined to become a worker bee, it will spend 21 days inside the cell before it reaches adulthood. After the first 3 days, the upright egg lays itself down as its membrane dissolves, revealing a tiny larva. The larva then goes through an astoundingly rapid 6-day growth period in which she must eat constantly. During this time, she sheds her skin 3 times, and her weight increases from about 0.1 mg to 150 mg. This would be like an 8-pound newborn baby growing to the weight of a 6-ton elephant in a matter of days!

To achieve this tremendous feat, the larva spends her first few days floating in a decadent pool of nutrient-rich jelly, moving only to eat. Her supply of brood food is frequently replenished by her adult caregivers. In her final days of feasting, pollen is added to her diet, and she is tended to thousands of times a day!

Throughout this period of development, because her digestive system has not fully formed, the larva is physically unable to release her waste. She is therefore prevented from contaminating her own food. Fortunately, once the gastro-Olympics has ended, her digestive tract quickly resolves itself so she can finally get some relief. At this point, the larva is still sealed inside her cell with a beeswax cap. She is no longer eating, but no one wants to be entombed with their own poop. So the

After a larva's six-day feeding frenzy, her cell will be capped over with wax.

larva spins a silken wallpaper to line the inside of her cell and cover the mess.

Now she is ready to change from larva to pupa. In time-lapse video, this process is strangely reminiscent of a popcorn kernel under pressure. Her outer layer gives way to a burst of fluffy white tissue that quickly resolves into the head, legs, abdomen, and thorax. At first she resembles a translucent, white alien shaped like a bee, but over the next 12 days her tissue reorganizes, her exoskeleton grows, and, finally, she sprouts hair. She emerges from her cell soft and covered in silvery hairs, but she soon hardens up and begins to work alongside her sisters.

Strange Genetics

Honey bees have baffling genetics. Most animals, like us, have two sets of gene-containing chromosomes: one from the mother's egg and the other from the father's sperm. Strangely, this is only true for the female bees in the hive. Drones are made from unfertilized eggs. While worker bees and queens receive two sets of 16 chromosomes, one from each parent, drones receive only one set and only from their mother (the queen). Yet the queen who produces them comes from a fertilized egg, so while drones do not have fathers, they do have grandfathers.

This is made possible by a special organ inside the queen called the spermatheca, which stores the semen from the males she has mated with. Essentially, the queen bee is a living sperm bank who can keep semen viable inside herself for years. If the queen wants to produce a worker bee or a replacement queen bee, she releases sperm along with her egg; but if she wants to produce a drone, she withholds the contents of her spermatheca and only releases the egg.

We don't yet know how the queen decides what kind of egg to lay, but at least part of the decision has to do with the size of the cell. Worker bees build two distinct cell sizes, one for workers and one slightly wider for drones. (A third, specialized cell for queens often hangs from the bottom of a comb and resembles an unshelled peanut.) It is believed that the queen can distinguish the type of cell by measuring its width with her foreleg before laying, but since she doesn't make use of every single cell, she must also consider the actual need for new bees in each caste. Since the worker bees build the cells and control the diets of the larvae (diets that can change workers into queens), it's probable that both the queen and the workers have some control over the population of each caste.

The Swarm: Child of the Superorganism

In the spring and summer, when nectar is plentiful, the colony swells fat with honeycomb, and the hive becomes pregnant with bees. Its population expands, and the hive bursts at the seams, sometimes spilling out of the entrance and clustering on the outside of the hive. The overcrowded colony is preparing to swarm.

In the final week before a swarm is born, queen cells are built along the edges of the combs. When the moment is right, the original queen and a large portion of honey-filled workers split off from the mother colony. They pour out of the hive and depart as an expansive, swirling cloud, hoping to establish themselves as a new colony.

Although the swarm has left its mother's body, it's still an embryo. It has all the potential for life, but it's not a fully developed superorganism. If it's going to survive, it must successfully implant itself in a new nesting location and grow its comb. Sadly, swarms have only a 25 percent survival rate; many don't make it.

The mother colony, too, is at risk after birthing a swarm. Not only did the creation of the swarm deplete its honey stores and worker population, but it also cost the colony its queen. If the mother colony fails to replace its queen, it will collapse. If the workers succeed, however, they gain a young queen with the power to invigorate the aging colony.

Despite the risks, some colonies send out multiple swarms in close succession, as if birthing twins or triplets. These second and third swarms are called afterswarms, and they are often smaller than the firstborn, or primary, swarm. Although the mother colony has a tougher recovery, multiple offspring mean its genetics have a greater chance of surviving and sending out more swarms the following year.

Catching Swarms

Many beekeepers jump at the chance to catch a swarm. Not only is it a way to gain a new colony with relatively little effort, but it's also fun! In some places swarms are so infrequent that beekeepers fight over the opportunity to retrieve them, while in other places swarms are so abundant that beekeepers will charge a fee to come get them.

In San Diego during the spring and summer, my phone rings day and night with bee removal requests, but it's always thrilling to drop what I am doing and rush out in pursuit of a swarm. I have removed swarms from car engines, boats, an aircraft carrier, a penguin exhibit, and even the stage of a Shakespeare play.

On some occasions, the swarms find me. At least once a year I watch a swarm fly over my backyard fence and make itself at home in a stack of unoccupied bee equipment. It's as if my backyard apiary has a large, flashing vacancy sign!

A swarm takes off in search of a new home.

CHAPTER 2

Busy as a Bee

How honey bees organize an amazing and ever-shifting workforce

A worker bee is not afraid to make a career change. Throughout her short life, she transitions through many roles and often performs more than one at a time. She remains in some positions for several days and is done with others in mere seconds. In general, younger bees work inside the hive, and, as they age, their jobs take them from the central brood nest to the periphery and, finally, to duties outside the hive. They are like our own teenagers, who start out with chores at home, then mow the neighbor's lawn for some extra cash before graduating to bagging groceries at the local supermarket.

Comb-building bees add lumps of beeswax to the edge of a newly constructed comb.

A beehive's entrance is crowded with guard bees, nectar receivers, and foragers.

A Shifting Career

A worker bee's job assignments follow a loose chronological order. Many tasks coincide with physiological changes in her body. But not every bee does every job. Certain bees might be genetically predisposed to certain tasks. The needs of the colony can also influence a worker's assignment. For example, a typical worker starts as a cleaner and then progresses to tending brood. From there she might build comb or help process nectar into honey, or she may do both these tasks at the same time.

Once a worker reaches middle age, she begins to forage. If the colony needs to increase its population, however, she may delay foraging in favor of tending brood for longer. If the scenario is reversed, she may progress more quickly through her house duties and begin foraging at a younger age. Each worker's career path is unique. Nonetheless, the following pages offer a basic outline of the common trajectory—and what is required of a bee for each task.

Cell Cleaning: Days 0–5

A worker's first job begins shortly after she climbs out of her cell. She is not fully developed, so her glands can't produce jelly, and she can't fly or even sting. She is therefore tasked with cleaning brood cells. Since the queen lays her eggs in large swaths, neighboring brood cells are close in age and will emerge at the same time. This convenient system not only places a young worker near an abundance of cells that need cleaning, it also means she need not wander far to begin her work. Each cell must be polished and made ready for the queen to lay another egg. Cleaners remove the remains of the cocoon and any excrement left by the previous inhabitant. The cell is then coated with a fresh layer of wax, sealing away anything that could harm a developing larva.

Nurse Bees: Days 2–11

Although she cannot have her own children, a nurse bee is a super-mom. Together with the other nurses, she feeds and cares for thousands of developing bees. The average larva is visited by nurse bees 13 thousand times a day! However, she's not only responsible for feeding the young; she must also feed adult members of the hive. Plus she often performs other tasks in addition to nursing, such as comb building.

A nurse bee produces a mixture of secretions called larval jelly from her mandibular and hypopharyngeal glands, which are only active in bees of nursing age and reach peak size between days 6 and 10. As a worker ages and moves on to other tasks, her glands shrink and take on different functions. Like mammary glands, they convert the food (bee bread) that a nurse bee eats into a kind of milk. These nurse bees consume more than 80 percent of all stored pollen in the hive. Think of a breastfeeding mom ordering two hamburgers at the drive-through!

Initially all larval castes are fed larval jelly, but, as with our own children, their nutritional needs change as they grow, and each caste demands a different diet. Like a mom with multiple toddlers, a nurse bee must tailor each lunch plate. Larval queens are fed only royal jelly throughout their development and do not receive any pollen. Meanwhile, worker and drone larvae are fed their own jellies, and, after four days, pollen and honey are added to their diets. The composition of all three diets changes daily, gradually containing more sugar and less protein.

When the time is right, the nurse bee seals each larva in its cell with a beeswax cap so it can begin to pupate. When the larvae emerge as adults, the nurse bee, like most moms, is there to feed them again. One-third to one-half of jelly produced by nurse bees is fed to adult bees in the hive.

A nurse bee feeds a young worker as she emerges from her cell.

Comb Building: Days 5–9

Young worker bees make comb from their own bodies. They excrete liquid beeswax that quickly hardens into tiny, translucent flakes from four paired glands on the underside of their abdomens and shape them into their trademark hexagonal forms. These golden, geometric structures are built from the top down and with surprising speed. A thumb-size comb can become hand-size in a matter of hours, and the bees often build two or three combs at a time. The comb builders form a kind of living lace. They hang with their outstretched limbs clasped together, and, like a magic trick, comb seems to suddenly appear from behind the curtain of bees.

In actuality, the wax flakes peek out from between abdominal segments. A spiny patch of hairs on the inside of the bees' back legs grips the flakes before passing them up to their mandibles. The young worker bee chews each flake, mixing it with mandibular gland secretions to make it more malleable, then compiles it with other flakes into a ridge where she sculpts offset hexagonal chambers on each side. As a comb builder works, she uses gravity-sensing hairs along her neck to ensure the vertical combs are plumb.

To our eyes, the combs look impossibly perfect, as if made by machine, but if we look closely, we will find that they are charmingly organic. Formed by thousands of tiny bee mandibles and tarsi, each cell is like the top of an artisanal ceramic coffee mug—textured and of varying heights, only an approximation of its intended shape.

Heater Bees: Days 2–27

Some bees like it hot—about 10 degrees hotter than other bees, to be specific. These special heater bees use rapid flight muscle contractions to raise their body temperatures and then strategically place themselves throughout the brood nest where the queen lays her eggs. Honey bee brood must be kept within a temperature range of 90 to 97°F (32 to

36°C), and pupae are especially sensitive to cold. All the bees in the brood nest contribute to maintaining the heat, their bodies covering the brood like a blanket, but heater bees seek out capped brood and transfer their heat to the pupae inside.

A heater bee can sometimes be recognized by her posture. She lowers herself to the cell with the intense concentration of a yogi, pressing her thorax against the cell cap, her antennae uncharacteristically still. Perhaps she is monitoring the rising temperature with them. Her heated thorax is perfectly sized to cover a single cell cap, but, whenever possible, she prefers to warm the brood from inside a cell. She climbs into unoccupied cells in the brood nest and radiates heat to neighboring cells. This heating method is efficient since it warms more than one bee at a time. For this reason, empty cells in the brood nest are likely deliberate. Amazingly, a heater bee will raise her temperature even higher based on the number of pupae that neighbor her cell. If she is next to only one pupa, she will average 91°F (33°C), but if she borders the maximum of six pupae, she averages 106°F (41°C).

The amount of energy a heater bee expends while heating is on par with strenuous flying. In a single 30-minute heating session, she spends the same amount of energy she would need for 12 foraging trips. Her energy expenditure is so high, she could keep a 20-watt lightbulb lit! Her job is so important that she is paired with special refueling bees. These helpful bees are like a NASCAR pit crew: They run back and forth from the periphery of the nest to bring honey to the heater bees when their gas tanks are low.

Ventilating: Days 8–25

Honey bees are living thermostats that are always working to maintain an optimal temperature for the hive. Not only do they heat their hives, but they also cool them. Around middle age, some worker bees turn themselves into tiny fans. These ventilator bees use their wings to cool

A ventilator bee helps to keep her hive cool by fanning her wings at the entrance.

the colony, evaporate water from honey, and reduce carbon dioxide and humidity levels inside the hive. This task is often what takes a worker bee outside her hive for the first time. On hot summer afternoons formations of fanners can be seen working at the hive's entrance. They stand a wing's width apart, their heads down and their abdomens curved. The combined power of their whirring wings is so strong that a beekeeper can feel the breeze on their face if they are close enough.

When the internal temperature approaches 99°F (37°C) the ventilators must act quickly, or their pupae may not develop properly. Even worse, if the temperature reaches 104°F (40°C), their beeswax combs will soften, and the heavy honeycombs could collapse. For the bees, a fallen honeycomb is the equivalent of an earthquake and a flood. Not only does their infrastructure crumble, but the honey that spills out into the hive will drown bees.

Each ventilator bee has a set temperature at which she begins fanning. If the temperature continues to climb despite the efforts of the first group, an increasing number of bees will join in on the fanning work and will not stop until the temperature has dropped below their own personal activation threshold. Fanning bees often work in groups and stand in lines to increase airflow along existing currents.

Undertakers: Days 7–20

Most bees die outside of the hive, but when they don't, their bodies must be removed to prevent the growth of harmful bacteria. Some middle-aged bees have the duty of removing these dead bees from the hive. Under normal conditions, undertaker bees are rare, representing just 1 to 2 percent of the colony. These dedicated bees are genetically predisposed to the job. They detect the dead and drag them out of the hive with their mandibles. Some toss them unceremoniously from the entrance, leaving a pile of dead below, while others go to great lengths to move them farther away from the hive.

It's not uncommon to see an undertaker struggling to lift a dead bee outside the hive. She may grapple with the corpse for several minutes before managing to fly off with it and drop it several yards away. In some instances, the undertakers bring to mind a scene from a certain Monty Python movie, for the bees are not always dead yet!

Guard Bees: Days 14–22

Given the delicious treasure inside, it's no wonder the beehive has guards at its entrance. Although we humans prize the honey, many species are even more interested in the protein-rich larvae. The colony faces myriad invaders: lines of marching ants, bold wasps, sly rodents, destructive bears, and even honey bees from neighboring colonies.

The guards are like bouncers, but instead of checking IDs they use scent to recognize their nestmates. At the entrance, guards can be seen rushing up to landing bees to inspect them with their antennae or a quick lick of the tongue before letting them pass. Sometimes they are quite rough and will grab a bee as she's landing to investigate her. If entry is refused, the guard will forcibly drag the rejected bee away, often with the help of other guards. The guard may even bite or sting if there's resistance, although they aren't very consistent.

Sometimes the guards let in bees who do not belong to their colony, and at other times they will refuse entry to those who do. We don't yet know exactly how a guard decides whether to admit a bee, but the intensity of guarding behavior seems to vary by foraging conditions and from colony to colony. One study found that guard bees rejected about 75 percent of non-nestmates when nectar was in short supply, but as conditions improved, they became more permissive and eventually let everybody in when nectar was abundant.

A guard bee rears up at the entrance to her hive in an effort to check incoming forager bees.

Perhaps guards are more protective during a nectar dearth because non-nestmates are likely there to steal honey. When nectar is plentiful, they become lax, perhaps because there are fewer robbing attempts. It could also be that at the height of the foraging season, guarding behavior is too disruptive to the busy foraging traffic.

Guard duty is not a popular job in the hive. Many bees never take on the role, and those who do often don't stay in the position for long. While most guards work for less than a day, some guard for as many as seven. All in all, there are usually fewer than one hundred bees guarding at any given time. If a guard bee senses a threat, however, she calls for backup (see page 110).

Nectar Receivers: Days 10–22

Honey bees have their groceries delivered. Nectar receivers wait at the hive's entrance for returning foragers who, like busy delivery drivers, are eager to get back to work. These foragers typically hand off their nectar to a receiving bee in under four minutes. The receivers lap up the offered nectar with their tongues and either distribute it to hungry bees throughout the hive or process it into honey to be stored.

While this is an efficient system to share the labor of making honey, it has a more complex purpose. The exchange between foragers and receivers is believed to be a way for foragers to assess the colony's food levels and to balance the workforce accordingly. If a forager cannot find a receiver bee to take her nectar, or if she must wait a long time for one to become available, she performs a dance to recruit more (see page 102).

Foraging: Days 18–Death

Beekeepers will tell you that watching the foragers at the hive's entrance is like staring into a campfire. The flurry of daytime activity is so constant, it's hypnotic. A healthy colony requires approximately 100 pounds of honey and 50 pounds of pollen per year to survive, so it dedicates one-third of its population to foraging. Tens of thousands of bees leave their hive each day to gather nectar, pollen, water, and plant resins.

When a bee approaches foraging age, she learns her surroundings by making short, looping flights around her hive. Incredibly, after just a few practice flights, she will embark on dangerous foraging expeditions that could take her up to 3.6 miles from her home. Given her body size, the distance a forager flies on a single trip is equivalent to a human traveling from Los Angeles to San Francisco. Since a forager flies 18 miles per hour, this is achieved in just 12 minutes.

Most bees begin their foraging career with nectar collection. A nectar forager flits between flowers, lapping up the sweet liquid with her tongue. Inside her body, her honey stomach expands like a water balloon with the collected nectar. It holds only a droplet, 60 mg, yet this is more than half her normal body weight! Although nectar collection may seem straightforward, it's actually complicated because every flower dispenses its nectar differently and at different times of day. As a forager grows more experienced, she's able to collect more nectar per trip and learn the time-sensitive patterns of the flowers around her hive.

A seasoned forager also becomes more likely to collect pollen. Depending on the type of flower she visits, she may collect nectar on one flight and pollen on another, or she may try to collect both at the same time. Since her flight is fueled by nectar, she must be careful about how far she travels to fill her honey stomach and pollen pants. She cannot spend more than she gathers. So not only does she manage to find her way home from great distances, but she also optimizes her flight paths to be as efficient as possible.

Some foragers only collect nectar and may never gather pollen, water, or propolis. Perhaps this is because the demand for nectar is constant, while the need for pollen, water, and propolis fluctuates by season. In spring, as the colony's population grows, pollen becomes

essential. Hungry nurse bees must eat lots of bee bread to produce larval jelly. In summer, as the days grow hotter, the colony may prioritize water collection to help cool the hive. In fall, propolis becomes a priority to ensure that the nesting cavity is well sealed and not too drafty for winter.

Although foragers are active from sunrise to sunset, most individuals don't work longer than four hours. A typical foragers makes 1 to 5 trips a day. As her experience grows, she may make as many as 10 trips, and some super-foragers have been recorded to make 17! Indeed, a small group of expert foragers appears to be responsible for much of the food collection. One study found that 20 percent of foragers performed 50 percent of their colony's total foraging trips, gathering more pollen and nectar than non-expert foragers. They were also more likely to collect pollen than the average forager, suggesting that pollen foraging may require more skill.

Water Collectors

Honey bees do not store much water inside their hive. Instead, a group of specialized foragers collects it as needed. These water foragers make up about 1 percent of the colony, but only some of them do the task full-time and never collect anything else. Water is required year-round because it's consumed by adult bees and larvae, but it's also used to control hive humidity in dry climates and hive temperature in hot weather.

When temperatures rise, the colony deploys additional water foragers who mob nearby water sources, sometimes preventing other thirsty animals from getting a drink. It's not uncommon for bees to commandeer birdbaths, dog bowls, or horse troughs on hot summer days. These foragers ferry water to their hive by filling their honey stomachs and may return to the water source several times to repeat the process. If needed, a water

A forager bee prepares to unload the bright yellow pollen she's collected into a nearby cell.

The Opportunistic Forager

Foragers are known to collect some unusual materials. I once witnessed several bees collecting the adhesive from an old piece of duct tape. My theory is that they used it in their propolis. It seems that duct tape really is good for everything! Duct tape foraging may be an anomaly, but it's not uncommon for foragers to mob a hummingbird feeder or even to rob soda from garbage cans during a nectar dearth. I once tasted some cherry-red honey from the combs of a colony located near a zoo. I was surprised by its unnatural cough syrup flavor but then realized they must have foraged on discarded Slurpees.

Foragers may also harvest the dust from chicken or bird feed at certain times of year. They seem to be able to extract some protein from it when pollen is in short supply. Similarly, I have seen them scraping powdery mildew off squash leaves in the garden and packing it onto their legs like pollen. The collection of various fungi may be another strategy to replace protein.

I've also noticed that, given the chance, bees will collect used beeswax from abandoned combs and equipment. If the wax is warmed by the sun, a forager can chew off chunks and press them onto her back legs to bring to the hive. Comb made from recycled wax often has a marbled look, with lots of tan-colored wax mixed in with the new, white wax. Although it's fun to watch, beekeepers who live in areas with high risks of American foulbrood (see page 157) have to be wary of letting their bees thrift old wax so that they don't spread disease.

forager can make as many as 50 water trips in a day. The total amount of water gathered per colony per day can be as much as 32 ounces!

The water is received by waiting bees near the hive's entrance. From there, the bees use it to create their own air conditioner. A thin layer of water is spread over sealed brood or on the rims of cells containing larvae and eggs. Then the adult bees fan their wings vigorously, creating air currents that evaporate the water and cool down the brood nest. Researchers have also observed that when hot weather persists, some bees—water collectors and others—transform themselves into living watercoolers. These watercooler bees fill their honey stomachs with water and station themselves around the hive, dispensing water to thirsty hivemates.

Scouts

Scout bees are an elite group of adventure-seeking foragers that represent 5 to 35 percent of a colony's foraging population. No matter how many profitable food sources a colony has found, a scout bee is constantly looking for a new one. She is like an explorer always in search of the next big discovery. When she finds it, she draws a treasure map for the other foragers to follow.

She directs them by performing an information-rich dance that communicates the approximate distance and location of flowers she has found (read more about bee dances on page 98). Typically, the recruited foragers will continue to visit the same site until the flowers are no longer in bloom or until they are directed to a better option. These recruits rely on the older, more experienced scouts to tell them where to go. A scout is vital to her colony's foraging operation. Her vigilant exploration enables the colony to adapt quickly to an ever-changing floral landscape.

When a colony is ready to swarm, a scout bee switches from searching for new food sources to house hunting. Her experience as forager makes her a good real estate agent because she is already familiar with the neighborhood. It's her job to find and evaluate potential nesting sites, and, if one of her sites is chosen, she will also guide the swarm to its new home.

Steering the Swarm

A honey bee swarm forms a large, swirling cloud in the sky. It looks chaotic, but if you watch closely, you can see that there is a pattern to it. The bees fly in all different directions, but they form an overlapping, crisscrossing pattern that looks almost like strands of a double helix. Amazingly, the mass of bees manages to stay together, sometimes traveling for miles to get to their new home, though only 5 percent of the bees in the swarm have been to the new nesting site and actually know where they are going. These scout bees steer the swarm by making what researchers have named "streaker flights." The scouts gather at the top center of the swarm mass and make quick, short flights in the direction of the new nesting site.

Although I have watched flying swarms many times, I am always delighted at the chance to see one again. I have even witnessed these streaking scouts with my own eyes, but it's still difficult to tell exactly how their flights motivate the other bees to travel in the right direction. Most of the bees look as if they are treading water, swimming lazy circles in the sky, yet somehow the cloud moves slowly toward its goal, the scout bees streaking through like shooting stars.

Sleep

Despite their reputation for industry, worker bees spend most of their time not working. They often wander the combs or stand completely still. They even nap! Of course, they could be gathering information as they patrol the combs, and when a bee is standing still she might be producing beeswax or brood food. These inactive bees may also serve as a kind of reserve force that can be sent in as backup when more bees are needed in a certain role. And we can hardly fault a bee who is napping. Worker bees sleep five to eight hours a day, and

A swarm of bees rests in a tree while the scout bees search for potential nesting sites.

when a bee is deprived of sleep, her ability to communicate and navigate becomes impaired.

Like us, bees experience different levels of sleep. They sleep lightly at first and then fall into a deeper sleep state. Some studies suggest that they may even have dreams! Bees also have a circadian rhythm. Young bees, who spend their time working inside the hive, follow an irregular pattern of sleep, alternating short stints of work with sleeping breaks throughout a 24-hour period. Very young bees engaged in cell cleaning are often found asleep inside an empty cell as if they'd dozed off in the middle of their cleaning project. As they get older, they are more likely to sleep on top of the combs in whatever part of the hive they have been working in. Beekeepers can recognize a sleeping bee by how still her antennae are—it's probably the only time they stop moving for a prolonged period. Like a tired puppy, the bee's head and antennae slowly droop down, and she may even flop onto her side as she falls deeper into sleep.

When a worker transitions to foraging, her sleeping pattern changes dramatically. Like a human with a nine-to-five job, she wakes up with the sun, works all day, and then sleeps through the night. Most foragers sleep inside the hive, but some nap inside flowers during the day. A group of scout bees will sometimes have a sleepover in a potential new hive site before the swarm moves in.

A tired forager rests on the comb inside her hive.

CHAPTER 3

What Bees Like

From flowers to fallen trees, things that attract honey bees and help them thrive

To truly understand the honey bee, we must highlight a few of her favorite things. Not only do her wants and needs give context to her behaviors, but they also help shape our world. When a honey bee favors a particular species of flower, she helps it thrive and spread across the landscape. Her preferences, however, are not idle whims. They have been honed through millions of years of survival and plant-pollinator interactions.

For the Love of Flowers

To say that honey bees like flowers is inadequate. The relationship between flowers and bees is intertwined and intimate. Scientists believe that flowers and bees evolved together in back-and-forth iterations that deepened their connection to each other and resulted in their mutual success. It's evident that honey bees are profoundly linked to the flowers they visit, but they don't interact with every kind of flower. They often give preferential treatment. Frustrated gardeners may notice that honey bees sometimes mob one plant while completely ignoring another. The reasons behind this behavior are more complex than most people realize. Flowers are not created equal.

Shape

A honey bee prefers flowers that are generously open, with welcoming petals extended like landing pads around a central fount of nectar and a visible ring of pollen. The pollen and nectar must be easy to access because a honey bee doesn't like to waste time. If a flower is too narrow to squeeze into or has too many fancy petals to sort through, she will find it less attractive.

Quantity

Honey bees like lavish gifts. To win a bee's favor, a flower must be rich in pollen and/or nectar, and the more it has, the better! Some flowers produce only one or the other, while some offer none at all. Outputs change not only by species but also by time of day and can even be affected by environmental factors like temperature, humidity, or carbon dioxide levels.

The bees also have an impact on how much pollen or nectar a flower contains. After a bee forages on a flower, the flower needs time to regenerate its gifts. Some species, like the blue star-shaped flowers of borage plants, are popular all day long because they refill themselves quickly.

Borage takes only two minutes to replace the nectar a bee has taken from it, while other flower types may take as long as 24 hours. A honey bee must constantly adapt to the changing nectar and pollen patterns in her environment. The object of her attention may change by the hour because she prioritizes the most profitable source. If a gardener only visited their peach trees in the late afternoon, they might mistake the lack of buzzing bees for disinterest, but if they returned in the morning, they might find their orchard filled with a riotous buzz.

When a bee comes back to the hive with a heavy load, other foragers pay close attention to her dance telling them where she's been so that they can forage there, too. You might imagine she has come from a vast wildflower field, but, in fact, most honey is made from trees and large shrubs. Honey bees favor large groupings of the same flower type because they only collect from one flower species per trip. A mature blooming tree that has thousands of flowers all in one place is a forager's jackpot.

Similarly, some species of flower are extra attractive because they grow in large spires or stalks and are not actually a single flower but many tiny florets. These individual florets may not produce more nectar than any other flower, but their proximity to each other makes them appealing. When flowers are close together, a bee can collect copious amounts of nectar without expending energy on flying. She moves only slightly as she works her way across the flower cluster, dipping her face into each teacup-like bloom to drink the nectar.

Quality

Despite her love of excess, a honey bee prefers quality over quantity. Nectar is mostly sugar and water, but the ratio of sugar to water varies, as does the blend of amino acids, carbohydrates, and other compounds found in nectar. When a honey bee has a variety of plant species to choose from, she tends to favor nectar with high sugar concentrations.

Some flowers produce thin nectar with more water than sugar. This blend appeals to birds and butterflies, but if its sugar level is below 20 percent, a honey bee won't bother with it. Her ideal nectar is between 35 and 65 percent sugar. Incredibly, some flowers know this and will increase the amount of sugar in their nectar if they sense the vibrations of a buzzing bee.

Pollen is similarly variable. Bees rely on pollen for protein and so tend to prefer varieties with higher protein levels. Depending on the species of plant, pollen can contain as little as 2.4 percent protein or as much as 60 percent. Yet pollen also contains important lipids, minerals, amino acids, and more. Honey bees cannot obtain all the nutrients necessary for their diet from a single pollen variety. They must forage on a variety of plants to maintain good health.

Medicine

For honey bees, food is medicine. Some pollen and nectar contain phytochemicals, antimicrobial ingredients that plants use for defense. Studies have shown that when eaten in moderation, these same phytochemicals can help protect honey bees from harmful bacteria, fungi, viruses, protozoa, and parasites. Researchers theorize that when bees have a diversity of plants to forage on, they naturally receive small, temporary doses of different phytochemicals. Since the bees adapt to an ever-changing pattern of blooming plants, they never forage on the same phytochemical for long, which prevents microbes and parasites from adapting to become resistant to the bees' defenses.

Honey bees may also seek out these antimicrobial foods deliberately when they are sick. One study found that nurse bees infected with the gut fungus *Nosema* chose honey with higher antibiotic activity over others. This suggests that honey bees may be able to use their stored honey and pollen as a kind of medicine cabinet.

Fine Weather

Honey bees are emblems of spring. They are most active in warm weather, when flowers are blooming and the days are long. A honey bee's ideal temperature range is what we all hope for on our beach vacations, between 72 and 77°F (22 and 25°C). Yet she also needs rain. Beekeepers will tell you that a dry spring means dry honeycombs. Without rain, the bees won't have enough flowers to forage on. If a spring is too wet, however, it can have the same result. The bees won't be able to leave their hives to forage at all if there is constant rain.

A bee's flying hours can also be restricted by low and high temperatures. A forager cannot take flight if it's colder than 50°F (10°C) or hotter than 100°F (38°C). These temperature swings can also impact their forage. Like the bees, most flowers do not thrive in low and high temperatures. The fluctuations can result in fewer flowers and less nectar per flower. Therefore, a honey bee's true weather preference is for balance and moderation.

Water Sources

Like all living things, honey bees and their larvae need to drink water to survive. They also use it to dilute crystallized honey, to maintain proper hive humidity, and to cool their hives during heat waves. New beekeepers are often puzzled when their bees fly right over the pristine water station set up for them in favor of a muddy, murky puddle. Honey bees seem to prefer dirty water and are often found collecting it from ponds, irrigation systems, pet dishes, and pools. They locate water by smell, so algae, moss, fish poop, chlorine, and other earthy odors attract them. Perhaps this method of locating water helps them to find nutrient-rich water, which often has plants growing in it and may have health benefits for the colony.

A honey bee also values water sources with safe places to land. Her ideal drinking fountain has ledges to perch on and floating plants that

Floating water plants provide safe landing pads for bees and add nutrients to the water.

act like rafts. She prefers still water over running water because there is less risk of being swept away. If she does fall in, she makes tiny waves by fanning her wings and is sometimes able to surf her way to safety. In the absence of ledges or plants to hang on to, she can easily drown.

The Dream Home

When a honey bee is house hunting, she views the world in terms of cavities. In wild spaces, honey bees typically nest in fallen tree trunks, hollows, or rocky caverns, but they've also adapted to urban spaces with the notion that any hollow object is a potential home. Bees will nest in mailboxes, outdoor furniture, garden tchotchkes—they'll even move into your attic! Regardless, each location is carefully chosen. Hundreds of scouts are deployed in search of a dream home with some specific criteria on their wish list.

When judging a potential nesting site, a scout bee may spend as long as 40 minutes on an inspection. The shape of the home is not important, but the swarm must have enough room to build a body of comb, so the scout bee carefully measures the space. To do so, she walks in a deliberate pattern over the interior walls and sometimes makes short flights between surfaces. Her ideal is thought to be 1.4 cubic feet (40 liters) of space, but depending on the options and genetics of the bees, a swarm may choose an even bigger home or settle for a smaller one.

A scout bee looks for a cavity that is well sealed, with good insulation. She prefers a small, easily defended entrance, but if other criteria are met, the swarm doesn't mind making modifications. If a nesting cavity gets too drafty, the colony can seal up any cracks and close off large entrances with propolis. A scout bee also finds previously occupied cavities attractive. Swarms tend to move into locations where bees have lived before and will reuse any comb or propolis left behind. Perhaps it's like moving into a furnished apartment!

If a scout cannot find an ideal home, just as she would forage on low-quality nectar, she will also settle for a less-than-ideal nesting cavity. Swarms sometimes move into strange places that may make it harder for them to survive. They may end up inside a flimsy cardboard box or between the folds of a beach umbrella. Some build their comb in the open air, hanging down from the underside of a hefty tree branch. In most climates these colonies will not make it through the winter, but in some mild climates they can thrive even without the protection of a cavity.

Sharing Walls

If you have the right outlook, you might find it endearing that honey bees sometimes choose to share our homes. They nestle themselves in spaces we aren't using: between studs, under eaves, or in an attic alcove. It seems the anatomy of a house provides many opportunities for nesting—all the bees need is a crack to enter. Once inside, they fill the space quickly with comb, and the hive may grow impressively large. When a wall is selected, they may construct nine-foot combs between studs that hang from ceiling to floor.

Despite their size, these extensive beehives aren't inherently damaging to the structure of the house. Unlike other animals, bees keep tidy nests: They don't defecate inside their hive, the honey is neatly stored, and they remove any dead bees. Conditions can sometimes conspire to collapse a comb. If the resulting honey spill overwhelms the bees, honey may seep through the walls and cause problems. Apart from the misfortune of an occasional honey leak, many colonies that choose to nest in our architecture thrive. They have shelter and ample space, and they may even benefit from home heating.

Microbiome

Living alongside the bees in the hive is a community of microbes that play a vital role in honey bee health and survival. Like a fingerprint, each colony has its own unique microbiome. These microbes contain a mix of organisms that are beneficial, harmless, or harmful to bees. The type and balance of microbes in the colony depends on its surrounding environment, the flowers its members visit, the amount of propolis it has, and the actions of its beekeeper (if any).

A portion of the microbes in the hive lives inside the bees' bodies and is passed from bee to bee during feeding and grooming. Studies have shown that it's critical for newly emerged worker bees to have contact with older workers. If a young bee is isolated from older bees, she does not develop fully and is visibly smaller. In addition those who do socialize with their older siblings receive helpful gut microbes that improve their digestion and strengthen their immune systems.

When a forager leaves her hive she brings back microbes on her body and in the food she's collected. Since each colony has distinct foraging preferences, this means that even neighboring colonies visit a different mix of plants and therefore acquire different microbes. It turns out that these microbes are the key to how nestmates recognize each other. They give the colony a unique scent that all the bees share, and this acts like a "members only" pass into the hive. To test this theory, researchers inoculated bees from the same colony with different microbes and found that afterward, they fought! They no longer recognized each other as being from the same hive, even though they shared genetics. For beekeepers, this may explain why young bees can be transferred and adopted by other colonies while older bees are rejected. Perhaps the young bees have not fully developed their microbial scent.

To some extent, honey bees can control the type of microbes that reside in their hive. They do so in part by maintaining temperature and humidity to a range where only certain microbes can thrive. They also constantly patrol the combs and walls of their cavity and keep their hive tidy. Propolis is employed as both a protective barrier and an antimicrobial salve that can be applied wherever outbreaks may occur. Even though propolis has strong antimicrobial properties, it doesn't seem to harm the microbes that benefit the bees. Colonies with higher propolis levels were actually found to have more beneficial microbes and fewer harmful ones than those with poor propolis levels.

Beekeepers can help their colonies thrive by roughing up the interior walls of their hives, which encourages the bees to layer more propolis. It should also be noted that some beekeeping practices can alter the microbiome in the hive, such as application of miticides or antibiotics, and perhaps even the practice of removing debris from the floor of the hive. In natural colonies, there is often a living floor below the combs made up of layers of fallen beeswax, pollen, and general debris that also contains a diversity of microbes and other insects. This living floor may be an important part of the colony, but more research needs to be done to understand exactly how.

A worker bee peeks over a barrier of propolis in the hive.

CHAPTER 4

The Bees' Knees

An up-close look at the honey bee's bodily wonders and how they connect to nature

The body of the honey bee is a marvel that's rarely appreciated. When depicted in popular culture, it is often reduced to a simple yellow puff with stripes. But a bee's body is luminous and layered—a dazzling complexity with purpose in every detail.

Pollen clings to every surface of this worker bee, thanks to her incredibly hairy body.

The Body of a Bee

On the most basic level, a honey bee body has four wings, six legs, and three main parts: the head, the thorax, and the abdomen. The head is the sensory hub. It encompasses the bee's eyes, antennae, and mouth, which are largely responsible for how the bee experiences the world. The thorax, with its connection to both legs and wings, is the bee's center of locomotion. The abdomen houses most of the bee's internal organs.

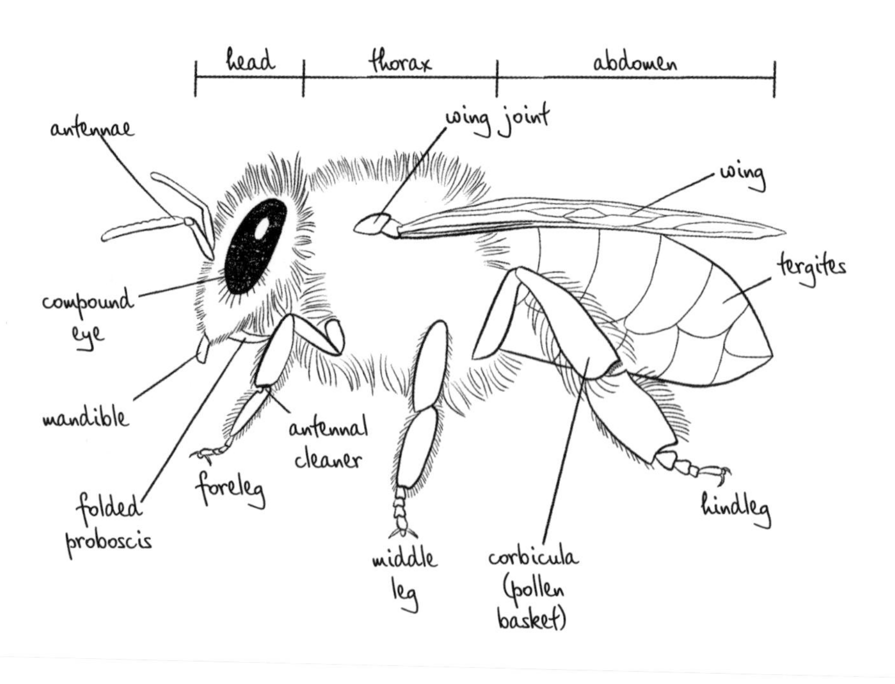

Exoskeleton

Each bee is sheathed in a shining exoskeleton. The hard, thin, overlapping plates protect the bee as well as give it form and function. Like our internal skeleton, the exoskeleton links all the bee's body parts, muscles, and connective tissues. On the abdomen, the plates together with patches of fine, shimmering hairs give the bee its color. It is not a uniform of yellow and black but a range of gold, amber, and umber. Up close, bees have the deep, marbled tones found in polished wood.

Hair

From antenna to hindleg, honey bees are covered in some three million hairs. When a young worker bee first emerges from her cell, her hair is matted and silvery. The damp swirls soon fluff up, giving the bee a golden luster. Not only does this fuzz make bees easier to love, it also helps them to collect pollen, protects them from debris, and keeps them warm. Some hairs even have a sensory function.

The hairs that are designed to trap pollen are called plumose hairs. They sprout from nearly every part of the bee's body—even the surface of her eyes! These hairs have a branched structure that is unique to bees and look like delicate, golden feathers under the microscope. After a bee visits a flower, the hairs become decidedly less elegant. They closely resemble the fingers of someone who's just eaten an entire bag of Cheetos.

Most bee hair is specialized for pollen collection, but some is designed to quickly release that same pollen when a bee grooms herself. To keep her sensory organs clean and working well, a bee will frequently swipe her hairy front legs over her head and body. The bristly leg hairs are designed to fling excess pollen away at high speed. Researchers found that a pollen-coated bee can clean herself in just three minutes, shedding up to 15 thousand pollen grains.

Antennae

When a hive is cracked open, curious antennae are the first to emerge. They swing back and forth eagerly, as if waving hello. These slender appendages move continuously as they smell, touch, taste, and listen to the world around them. A bee's antennae also detect flight speed, temperature, carbon dioxide, humidity, gravity, and possibly more!

Each sensory wand is connected to the bee's head by a ball-and-socket joint that allows it to swivel and point independently. Although an antenna has 12 segments, it looks like two long sections separated by an elbow joint. The antenna's magical powers of perception are made possible by thousands of tiny sensory structures. Most are pores, plates, pegs, and hairs, but one surprising sensor, which detects temperature and humidity, is shaped like a mushroom. The pores and plates are generally used for smell, while hairs and pegs have a tactile or taste function. Some hairs measure flight speed by feeling the pressure of the wind as the bee flies. Other hairs belonging to the Johnston's organ inside the antennae are sensitive to sound (more on page 92).

The information that all these antennal structures collect travels directly to the brain through a nerve trunk inside the antennae.

Eyes

The eyes of a honey bee are warm and chocolatey. They curve around the sides of the bee's head and seem to gaze from every angle—like the Mona Lisa. This is because their surface is made up of thousands of tiny hexagonal lenses. Each lens sits atop its own cone-shaped organ called an ommatidium, which extends back toward the brain. Every ommatidium captures an image from a slightly different vantage

Honey bee hairs range from long and feathered to coarse and spiky.

point, and the brain combines them into one cohesive picture. The result is that bees see the world as a kind of pointillist painting.

Honey bee eyes are textured with the same hexagonal pattern of their combs, but they don't look geometric. The pattern is so small, it blurs into a soft, velvety texture that reflects light diffusely. Yet their eyes are not without sparkle. Each one is crowned with glittery, golden lashes that spring from the surface of the eye as well as line the edges. Primarily they protect the eyes from pollen. The long hairs hold pollen grains away from the eyes, where they can be easily brushed away. Some eye hair also has a sensory function that helps bees to fly by detecting wind speed and direction.

Although their two large compound eyes are the most visible, honey bees actually have not two, but five eyes. Nestled in the fuzz, on what could be called the bee's forehead, is a set of three simple, single-lens eyes called ocelli. The ocelli are arranged in a triangular pattern. They are small, round, and easily missed. Indeed, there is something secret about them because we don't fully understand their purpose. It is believed that the ocelli do not "see" images; they only perceive light intensity. This function may help foragers know when to begin foraging in the morning and when to return home in the evening. Ocelli may also help foragers to keep their bodies level during flight based on the brightness of the sky and could also play a role in navigation.

Mouthparts

The mouthparts of the honey bee are wonderfully complex and versatile. Although they have many components, they can be generalized into two main structures: a set of mandibles and a tongue (also called a proboscis).

The mandibles hinge from the cheeks of the bee and, when closed, form the triangular tip of the bee's heart-shaped face. The smooth, amber paddles are reminiscent of tortoiseshell sunglasses. Long, gold hairs line the outer sides, while short sensory hairs frame the sharp inner edge. The mandibles move from side to side and are capable of biting, chewing, shaping, and carrying. They are used in nearly every aspect of colony life.

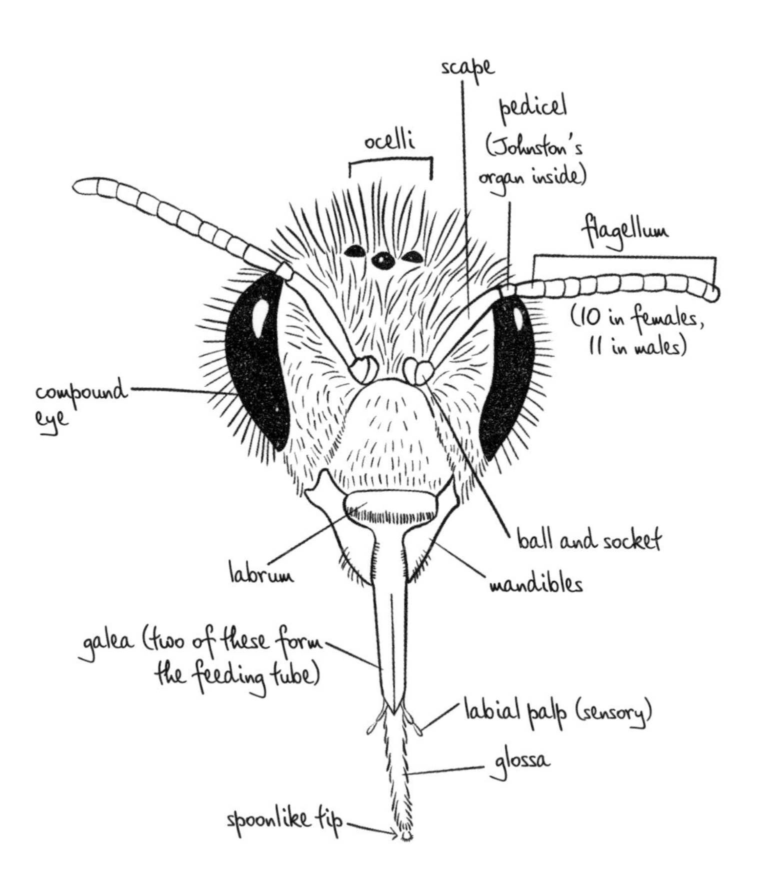

When a worker bee has finished developing and is ready to emerge from her sealed cell, it is her mandibles that free her. Just below the beeswax cap, she turns in circles, cutting along the edge, like a can opener, until she can exit. Later she may use her mandibles to shape beeswax into comb or to carry debris out of the hive. Her mandibles can serve as a spoon for feeding young larvae or as a brush for grooming nestmates. When invaders enter the hive, she wields her mandibles! She bites, chews, and drags out the intruders. As a forager she may chomp down on flower anthers when collecting pollen or gnaw at hardened sap when gathering ingredients for propolis. Her mandibles are also used to steady her tongue as she collects nectar or water. And when she reaches the end of her life, she will be carried from the hive in the mandibles of an undertaker bee.

The tongue is a feathery, vermillion apparatus with a spoonlike tip that reaches out from between the mandibles. It is sheathed by a shiny feeding tube and flanked by two slender sensory palps that have the ability to feel and taste. When a bee is foraging, the tongue extends beyond the feeding tube in a rapid dipping motion that propels nectar upward. Submerged in liquid, with its many hairs fanned out, the tongue resembles a kind of crimson seaweed. Its hairs help to trap and pull liquids into the feeding tube. Bees use their tongue to collect both nectar and water. They can even adjust the spacing of their tongue hairs to match the viscosity of the fluids they collect.

Back at the hive, the tongue is also used to distribute water and nectar. Returning foragers present droplets to receiver bees, who extend their own tongues to accept. These tongue-to-tongue exchanges can often be observed from outside the hive at the entrance.

A worker bee uses her proboscis to lap up spilled honey.

Wings

Sheer, shapely, and sparkled with light, the delicate wings of a bee are artfully engineered. They are not true appendages but an extension of the bee's exoskeleton. The thin membranes are strengthened by a network of amber veins. They divide the wing into shimmering opalescent sections, resembling a delicate work of stained glass. These veins give the wings structure and also carry blood and contain breathing tubes and nerves. The veins follow a characteristic pattern that varies slightly by breed and can be used to trace genetic ancestry. Bee wings may resemble glass, but they aren't smooth. Each rippled section catches light like a lake at sunrise. Yet beneath the champagne brilliance, short, spiny hairs dot the surface. These curious quills may serve as flight sensors.

Although a bee has four individual wings, each set functions as one during flight. The larger forewing connects to the hind wing via a row of tiny hooks and loops along their edges. Much of the thorax is taken up with flight muscles. These powerful muscles allow the wings to move up and down, forward and backward, and even to twist! When in flight, honey bee wings do not simply flap like those of a bird but tilt and turn as if paddling through the air. From the side, the tip of the wing traces the shape of a figure eight.

Speeding along at an average of 15 miles per hour, the wings of the bee are an iridescent blur. They beat at a miraculous speed of 230 beats per second. This is faster than a bee's brain can deliver the message to move via nerve impulse. Yet it is possible because bees store energy in their muscles, enabling them to beat their wings more than once each time they receive an impulse. When watched in slow motion, honey bees seem barely able to control the awesome power of their own wings. They wobble wildly with their legs dangling, sometimes colliding with each other at the hive entrance.

Wings aren't only for flying—they also function like fans. They are used to dehydrate nectar into honey, to cool the hive in hot weather, and to spread scent signals throughout the colony. It is flying, though, that eventually takes its toll on the wings. Some studies have shown that wings can carry the bee no more than five hundred miles in her lifetime. With use, they become tattered and torn, their performance gradually declining, until they finally give out.

Legs

Three pairs of legs unfold from the honey bee thorax like tools from a Swiss Army knife. Each has five polished, oblong segments that culminate in clawed feet, but the legs are not identical. They feature an assortment of notches, dips, hairs, sensors, and glands that help the bee with specific tasks. With her legs, a bee can climb up combs, clean herself, and collect pollen. She also uses her legs to taste, hear, and communicate.

The front legs have an abundance of extra spiny hairs that groom the head. Foragers swipe them across their eyes like windshield wipers or bring them together to clean their tongues, brushing the front and back simultaneously, a maneuver that also coats the forelegs with sticky nectar so that they work like wet wipes. When an antenna needs cleaning, the bee clamps it in the crook of her foreleg and pulls it through a circular notch lined with stiff bristles. This special antenna cleaner is found only on the front legs, and its diameter perfectly matches the antenna.

The middle legs also help to clean. When a bee grooms herself thoroughly, she flings them upward and rakes them across her back as if she were doing yoga. Despite her efforts there is an unreachable section along the top that sometimes accumulates pollen and resembles a racing stripe until it is cleaned by a sister bee. The middle legs also feature a mysterious spike whose purpose is unknown.

The hindlegs are highly specialized for pollen collection. A successful forager can hold so much pollen on her back legs that it triples her width. She lumbers from flower to flower like a semi with a "Wide Load" sign, adding more and more pollen at each stop. The packing process begins when the bee wades into the first flower of a foraging trip. The powdery pollen grains cling to her body hair. She gathers them by grooming herself with her front legs, adding nectar, and forming the pollen into a doughy mass.

The pollen is passed to the inside of each back leg and held in place by nine neatly organized rows of downward-pointing bristles dubbed

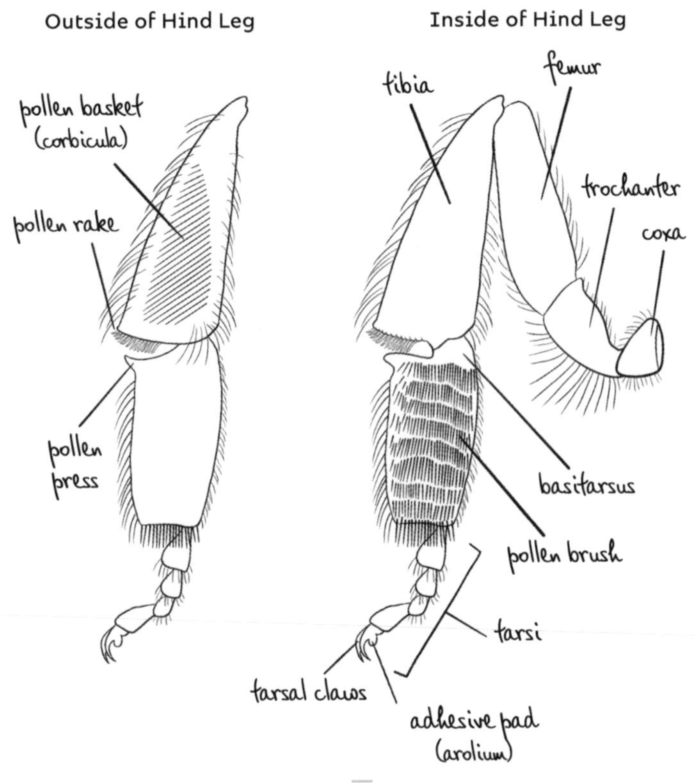

the pollen brush. Just below the pollen brush sits an important scraping tool, a singular row of extra thick bristles called the pollen rake. When the two back legs are rubbed together (brush to brush) the pollen mass is scraped off by the pollen rake on the opposite leg and forced into a gap, called the pollen press, between the segments of the leg. When the leg is straightened, the pollen is squeezed through the press and forced up into a basketlike indent on the outside of the leg. The corbicula, or pollen basket, is lined with long, curved hairs that help to hold the pollen in place. Each time the bee repeats this process, she adds more pollen to her baskets. Eventually each leg forms a huge, rounded mound that seems to teeter impossibly like an ambitious scoop of ice cream atop its cone. When she can carry no more, she makes her way back to the hive to unload.

In shape, honey bee feet resemble fleurs-de-lis. Although they are dainty, they are designed for utility. Four curved claws work like grappling hooks and can grip textured surfaces, but they are also used to taste. When visiting flowers, foragers dip their "toes" into the nectar to measure the sugar level. This helps them decide if the nectar is worth collecting. Between the claws, an adhesive pad expands and clings to smooth surfaces like a suction cup. On bumpy, abrasive terrain, the soft pad is tucked away and protected, but on slippery surfaces the claws are shunted to the sides, and the pad takes over completely. A nearby gland secretes an oily liquid that increases surface tension and helps the pads to stick, even on vertical planes. The liquid, called the footprint pheromone, also leaves behind a scent trail that guides other bees both to the hive entrance and to foraging sites.

Stinger

Hidden inside the abdomen is the honey bee's stinger. Once used by ancient bees for laying eggs, the modified ovipositor found only in female bees now dispatches venom instead. Like an elaborate gadget from a James Bond movie, the three-part stinger is made up of two barbed lancets and a stylus that connects to the venom bulb. Together they form a syringe-like sting shaft that injects venom. The lancets move like a boxer delivering a one-two punch. Paired muscles drive them forward in quick, alternating jabs. When the first lancet punctures its victim, the curved barbs anchor it in place. They are sharpest near the lancet's tip, presumably for ease of entry, but, viciously, the barbs increase in size as they move up the shaft. Once anchored, the stinger can only move forward. As one lancet retracts, it advances the other deeper and pulls the needle tip of the stylus closer with every movement.

Venom, delivered through the sting shaft, is a complex cocktail with ingredients to punish a wide variety of predators and honey thieves. It contains 63 different compounds, but melittin is the most abundant and the most toxic. It makes up 40 to 60 percent of the whole venom and is responsible for most of the pain and tissue damage. Venom composition changes throughout a bee's life, with melittin reaching its peak concentration at four weeks of age. This means that a bee's venom has the greatest potential for inflicting pain around the time that she is working as a guard or forager bee. A bee also produces more venom during the summer months when the colony is likely to have lots of honey in need of protection.

After delivering her sting, the bee flies off, and if her stinger has gone deep enough for the barbs to take hold, it is torn from her. The stinger and a pulsating venom sac continue to pump in venom even without the bee attached. Ninety percent of the venom is delivered in the first twenty seconds, and it takes only one minute for the venom sac to empty. The now stingless bee may live for hours after rupturing her abdomen, then eventually die. The time preceding her death is not wasted. She uses it to intimidate her victim by furiously buzzing and burrowing as if she will sting again. The sting site is also marked with

The Truth about Stings

I admire the success of the honey bee sting. The bee who stings is sacrificed to protect her colony, but her human victim is unlikely to die. The pain of her venom is so memorable, however, that a healthy fear of it is part of our collective wisdom. For most of us, though, a sting is not lethal. In fact, the average adult can withstand more than one thousand stings. Currently only 5 percent of the population is allergic to honey bee venom, but many people mistakenly believe themselves to be allergic, despite having normal side effects.

Every sting is a little different, but generally it starts with a sharp, white-hot pain that radiates outward. The intense pain will lessen within a minute but may continue to throb for as long as 15 minutes. It's normal for the sting site to be tender, swollen, and hot to the touch. This can last for days. The degree of swelling is often alarming: A sting on the finger may transform your entire hand into a puffy Muppet paw. The swelling might even extend as far as your elbow, and you are likely to experience itching as it heals. This isn't considered an allergic reaction, because the effects of the venom occur close to the sting site.

A true allergic reaction is any reaction not local to the area where you were stung. A normal reaction to a sting on the finger should not include swollen legs, for example. Nor should any sting include a full-body reaction, such as breaking out in hives or feeling lightheaded. Some allergic reactions manifest as nausea, cramps, diarrhea, or vomiting. Others are anaphylactic and include swelling of the throat or shortness of breath. Thankfully anaphylactic reactions are rare and have occurred in less than 1 percent of the population. Anaphylaxis is usually reversible if treated quickly with epinephrine (adrenaline).

It should be noted that even a normal sting reaction could become infected and result in a trip to the doctor. An infected sting is not an indication of an allergy to venom. It's only a bit of bad luck.

an alarm pheromone. If the sting victim is near a hive, the alarm pheromone is the bee version of a call for backup. Other guard bees will seek to defend their hive against the perceived threat by chasing, buzzing, and even stinging. Experienced beekeepers know to quickly smoke any bee stings they receive, or else they may receive even more.

Sometimes a bee can sting without losing her stinger or her life. It depends on how thick-skinned the victim is and how deeply the stinger penetrates. A bee can often sting through the thin exoskeleton of other insects and keep her stinger intact, but when defending her hive from a thick-skinned mammal, she may need to drive the stinger deeper. Beekeepers can attest to the varying pain levels of bee stings. Younger bees sometimes deliver shallow, relatively painless stings, seemingly without damaging themselves. Other times, an enraged guard will sting so deeply that the area appears bruised and remains purple, swollen, and throbbing for days.

Internal Systems

Inside the honey bee there are familiar features but with notable differences from our own anatomy. A honey bee has a brain, a heart, a stomach, a small intestine, a rectum, and many glands, muscles, and nerves—but she lacks lungs. Air is taken in through tiny holes called spiracles along the sides of her body. The spiracles are attached to a complex system of tracheal tubes and air sacs that help deliver oxygen where it is needed. A resting bee might pump her abdomen rhythmically, but she isn't preparing to sting—she's only catching her breath! A bee must move her entire abdomen to breathe.

A bee's circulatory system is also different. Blood (hemolymph) does not flow through a network of veins like ours does; instead, it circulates freely throughout the body cavity. It's pumped by a long, tubular heart that runs along the top of the bee's abdomen. Small holes along the heart's sides pull in the blood and transport it to the head, where it starts its cycle anew. Instead of carrying oxygen, the blood transports nutrients that it absorbs in the abdomen through food digestion just before reentering the heart.

Food travels through an esophagus before entering the honey stomach in the bee's abdomen. This "stomach" is not accurately named because no digestion takes place in it (there is a separate, actual stomach for that). The honey stomach is more like an internal shopping bag designed to carry nectar and water back to the hive. An empty honey stomach is like a limp balloon, but, as the bee fills it, it expands to ten times its size and will take up nearly half the bee's abdomen. A bee with a full honey stomach is visibly fat, and her engorged abdomen seems almost to glow from the nectar within.

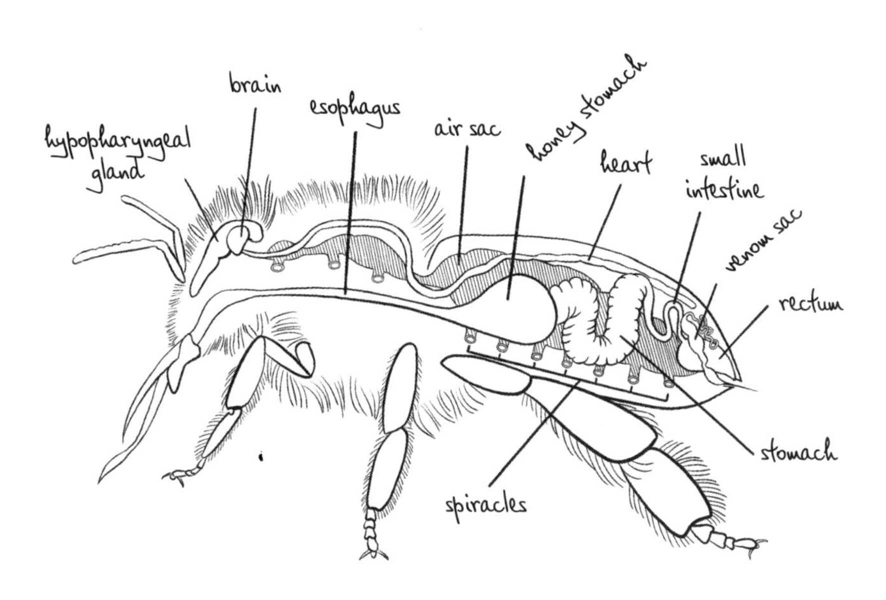

Even Bees Poo

Worker bees follow strict rules of hygiene and don't poop inside their hive. Instead, they fly out and away from the nest to relieve themselves. They are so dedicated to this practice that they will hold their waste over long periods if necessary. In most regions, honey bees are hivebound through winter and will not poop for many weeks or even months. They wait for warm winter days to get their relief, only taking these cleansing flights when temperatures rise to 50°F (10°C). New beekeepers are sometimes alarmed by the sudden appearance of yellow, mustardlike streaks in the snow around their hives, but this is actually a good sign. It means that the bees finally got some relief.

While the worker bees are careful to poop outside the hive, the other members of the colony are less so. As one might expect, the queen, who rarely leaves the hive, poops where she pleases, and workers clean up the mess. Drones may poop in or out of the hive since they do leave to mate, but workers are quick to clean up their accidents.

Although the subject of bee poop may seem esoteric, anyone who has ever washed a car has probably noticed stubborn yellow deposits dotting their windshield. Impervious to soap and scrubbing, these dots of bee poop can only be removed by scraping with a fingernail or other tool.

When a bee needs to eat, a valve that separates her honey stomach from her real stomach opens. Food passes into the stomach and is digested. The stomach lining is semipermeable so that nutrients from the food can be absorbed straight into the blood. From there the food travels through the small intestine (ileum), where water, salt, and other substances are separated from the waste and absorbed. The waste continues to the rectum and is excreted, a soupy yellow mixture of unused pollen components.

A Queenly Physique

The queen's most recognizable feature is her large abdomen. As she glides across the comb, her long body is reminiscent of a royal bustle. It trails behind her in a smooth, graceful movement that only she possesses. Although at first glance her abdomen is roughly the same shape as a worker bee's, it's actually slightly wider at the base and longer overall. When looked at from above, the queen's abdomen has six overlapping segments just like a worker bee's, but they don't overlap as much, so more of each segment is visible. This is because her abdomen houses large ovaries that swell even larger after she mates and take up nearly two-thirds of the space inside. Along with all the necessary organs, the queen also has a special sperm-storing sac called a spermatheca, which she fills during her mating flight (see page 129). Her wings, too, are marginally larger than the workers', but her abdomen is so long that her wings appear much smaller. A queen has three visible abdominal segments past her wing tips, while a worker has none; the worker's wings cover her entire abdomen.

The queen's shiny, black thorax is another prominent feature. It's lack of hair makes it eye-catching, and, again, it's just a bit bigger than the worker bee's. Beekeepers may be interested to know that this slight size difference is what prevents the queen from passing through a queen excluder.

The queen's long, golden legs share the same basic structure as the workers', but they do not have the fancy pollen-gathering features, since the queen doesn't collect pollen. However, she does still have plumose hairs. The queen's head is also very similar to the worker bees', with only minor size differences in her eyes, ocelli, antennae, and mandibles. Her final distinguishing feature is her smooth stinger. Although she rarely uses it, its lack of barbs allows her to sting repeatedly, like a wasp.

Drone Bods

Drones are markedly larger than worker bees and are often mistaken for queens simply because of their size. However, they are much squarer in shape and have a generous layer of manly fuzz that, unlike the queen, covers the top of their thorax. Their big, powerful wings stretch past the tips of their abdomens, and a massive set of round eyes crowds the head. In fact their eyes are so large, they have nearly double the number of lenses a worker has, which surely helps them on their mission to mate with a queen.

The drone's antennae are also larger and have an extra segment, 13 to to the worker's 12. The additional surface area means he has space for nearly 16,000 olfactory sensors, an incredible number when compared to the worker's 2,700. Like the queen, he has all the standard organs, but much of his abdomen is dedicated to reproductive parts. However, since male bees never had the ability to lay eggs, none of them (even those of other bee species) have the stinger that evolved from that ancient organ.

Worker bees gather around their queen.

CHAPTER 5

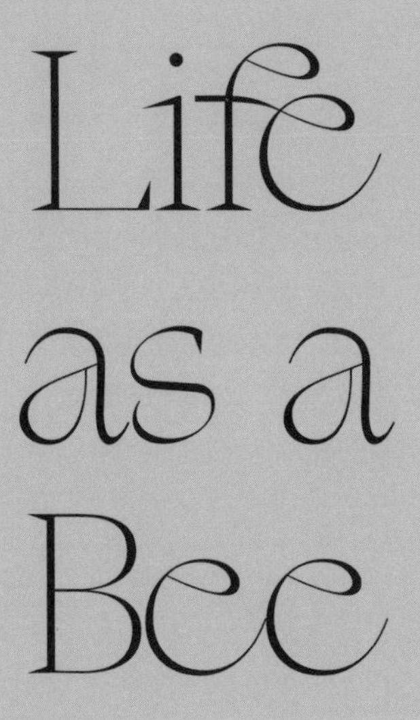

Life as a Bee

How a honey bee senses and experiences the world

Hasn't every naturalist wondered what it's like to roll in a flower and drink its sweet nectar?

Imagine life as a bee. To grow up in a warm, cloistered chamber surrounded by sisters and then to take flight into the bright unknown. Your antennae guide you through blasting wind and across unfamiliar landscapes. You must rely on your memory, sight, and sense of smell to take you to the flowers you seek and then home again.

A forager bee buzzes through a patch of blooming clover.

The Brain of a Bee

A honey bee brain is the size of a sesame seed, but it's shaped like a video game controller. It stretches between her two large eyes, the optical lobes widening at each end. It is 20 thousand times smaller than our own brain, yet it is capable of complex learning we normally only associate with vertebrates. Despite its small size, the honey bee brain is ten times denser than a typical mammal's, with nearly one million neurons packed into it. It has dedicated lobes that collect information from the bee's sensory parts, like its eyes and antennae. Another section of the brain, the mushroom body, is responsible for learning and memory.

Learning

Honey bees have an astonishing ability to learn. They can recognize and remember colors, shapes, patterns, and scents. They can count, add and subtract, sequence, and combine concepts that they learn. Scientists test their abilities by offering the bees sugar water as a reward. They often present them with options, such as two colors or two shapes, and the bees quickly learn which one results in a sweet treat. Yet bees are capable of processing incredibly complex options. One study found that they could correctly identify different art styles. The bees were able to discern Picasso from Monet and, when presented with a new painting, were able to generalize whether it was a cubist or impressionist work.

Bees understand abstract concepts, like when an object is the same as or different from another object. In one study, they were shown a symbol as they entered a Y-shaped maze. When the bees reached the fork, researchers were able to teach them to select the path marked with either that same symbol or one that differed from it. Bees have also been credited with understanding the concept of zero, an ability few other animals possess and one that even human children struggle with. In a study in which bees were taught to identify images that had the fewest number of dots, they correctly chose a blank image over one with dots.

Ironically, one of the most impressive aspects of honey bee learning might be their ability to admit when they don't know the answer to a problem. This kind of self-awareness is attributed only to humans, and there is much debate about whether animals are capable of it. Researchers explored this idea with a series of trials that rewarded bees for right answers and punished them for wrong answers, but they also gave the bees the option of not answering. As the trials grew in difficulty, the number of bees who chose not to answer increased. Was this their way of declaring, "I don't know"?

As impressive as these findings are, honey bees are probably not sitting around their hives solving math problems. Their abilities stem from their adaptability. Since the floral landscapes are always changing, bees have become very good problem-solvers. They seem to be able to adapt to almost any circumstance, especially when motivated by a sweet reward.

Memory

Honey bees learn and form memories through experiences. When a forager discovers a patch of nectar-rich flowers, she memorizes their scent, color, and location. She can remember route details, such as landmarks, up to six miles from her hive. She stores these visual landmark memories in sequence for both her outbound and inbound journeys. She can also memorize the best time to visit a flower. Many flowers only produce nectar at certain times of day. A savvy forager learns the flowers' schedule and will show up at the correct time until the blooms are spent.

Honey bees have short-term memories that last seconds to hours and long-term memories that last days, weeks, or even months. Some long-term memories are dependent on the bee's lifespan. A newly minted summer forager may retain her memories for the remainder of her life (about two weeks), but an overwintering bee may keep hers for months. It's not known exactly when a honey bee begins to form memories and how long they might last. One study found that scent memories established by young bees who had only emerged from their cells a few days before were recalled once these bees became foragers 9 to 12 days later.

Vision

When a honey bee peers back at you, she sees your face as a pixelated pattern. Each pixel is captured by an individual ommatidium, one of the cone-shaped optical units that make up the compound eyes. Her mosaic-like vision lacks details, but she sees well enough to differentiate one human face from another. If you were to wave hello to her, your hand might come more clearly into focus. This is because a moving object engages more ommatidia and allows them to collect more pixels. A bee's eyes work like a slow-motion camera, collecting thousands of photos and combining them. It is the opposite of what motion does to our vision. If a bee were riding in the back seat of a car, the roadside flowers wouldn't blur as the car picked up speed; they would become sharper.

Honey bees also see more clearly when they themselves are in motion, which may explain why they sometimes appear to be scanning beekeepers and other objects of interest, flying in slow zigzags as they investigate us from head to toe. Could it be that they are trying to see us more clearly? Still, honey bees don't see details as well as we do. A bee can clearly see something the size of an olive from two feet away, but anything smaller begins to blur.

Honey bees see in color, but they lose this ability at high speed. The world becomes black and white at 18 miles per hour (30 kmph). This may be because it takes more energy to see in color, and refined vision

Stranger Danger

Honey bees' talent for identifying and remembering different flower types means that they also have the ability to recognize human faces, but we don't know if they put it to use outside of a laboratory setting. However, I sometimes suspect that honey bees know their beekeeper and are at times wary of strangers.

My honey bees seem to recognize me as an individual and therefore know when there is a new person in the apiary. On several occasions, I have seen colonies that are always docile with me be less so with others. They sometimes investigate new people with scanning flights or even chase them away threateningly. I have also inherited colonies that have been neglected and gone for years without regular inspections. These colonies are often hostile at first but will settle down after a few months of regular hive checks.

A few years ago, I was discussing this curiosity with a beekeeper I met at a conference where I was speaking. He was in his nineties and had kept bees for many decades. He told me that his mentor had once advised him to hang his dirty clothing on a post in front of any defensive colonies so that they might get used to his smell and tolerate him better. It leaves me wondering: Could honey bee colonies develop a trust in their beekeepers as other animals do with their human caretakers?

is not needed when traveling. Like us, honey bees have trichromatic color vision, but while we see red, yellow, and blue and their combinations, they see green, blue, and ultraviolet and their combinations. Additionally, honey bees cannot see the color red at all; instead they see red as black. Nonetheless they are privy to a secret world of color that we can only try to imagine through technology. Special cameras have revealed that many flowers have bold UV patterns that are invisible to us but attract bees. These UV markings act like neon signs advertising the nectar within. As the bee flies overhead, she doesn't see a simple yellow flower, but a sparkling, multicolored bullseye.

As the honey bee flies, she uses the direction of light waves to help her navigate. Light waves move in all directions, crisscrossing the sky with lines that are not visible to us. When light hits certain surfaces like clouds or water, those surfaces act like filters, only letting through light that is moving in the same direction. The resulting streams of light are what we call polarized light, and these pathways of light help guide the bee as she flies.

A Comedy of Errors

Although it often seems like honey bees are perfectly designed, they are not impervious to error. I once witnessed a hilarious phenomenon of navigation after installing a new metal roof on one of my hives. The pristine, extra shiny roof was reflecting so much light that it was confusing the bees who flew above it. Each bee who passed would suddenly twist in midair and then land with an audible *thunk* on her back. As I watched, dozens of bees crash-landed one after the other, each flailing wildly for several seconds before finally righting herself. It was as if the light reflecting off the roof was somehow brighter than the actual sun, and it tricked the bees into thinking they should be flying upside down!

After several minutes, I decided to help the bees out by covering the roof with cardboard. The crash-landings stopped, and I never witnessed this oddity again. Luckily for the bees, it seems that the light conditions that trigger upside-down flying are rare.

A worker bee in flight relies on her senses to guide her.

A bee has specialized cells in her eyes that allow her to see light in different ways. Some cells detect all the light waves at once, while others detect only certain planes of polarized light. It's as if she is wearing many tiny pairs of sunglasses and can switch to the ones that function best according to the changing light conditions. Thus her three simple eyes, the ocelli, together with an upper rim of specialized ommatidia, detect the pattern of polarized light across the sky. It is a celestial compass that guides her to and from her hive, even on a cloudy day.

Smell

The nose of a honey bee reaches out from her face and swings about in search of scents—it is the bee's antennae. These curious appendages are covered with thousands of sensors, but more than half of them are used for smell. The oval-shaped, platelike sensilla perceive scent molecules in the air and relay that information to the brain, where it is processed, stored, and even recalled as scent memories. A honey bee can be trained to recognize and remember odors, often in a single session. She can also remember a sequence of smells in the correct order. Although researchers have trained bees to detect bombs, cancer, and COVID-19, their refined olfactory capabilities exist because they are crucial to colony survival.

While foraging, a honey bee encounters incredibly complex floral scents, each one with potentially hundreds of chemical components. These odors are extremely changeable. They differ by floral species, time of day, pollination status, nectar levels, and the age of the flower. A honey bee can smell these differences and uses them to decide which flowers are rich in nectar and therefore worth visiting. As she becomes more experienced, she learns to generalize and will sniff out flowers with scents that are similar to ones she has visited before. If they share some of the same scent components, it's likely that they are similarly bountiful in nectar.

A bee's sense of smell, like ours, is closely linked to memory. When she gets a whiff of a flower she's dined on before, it triggers a visual memory that allows her to navigate back to the place of the meal—like how the smell of pepperoni makes you want to visit your favorite pizza joint again. She can even get nostalgic for childhood meals and will show a preference for flowers whose nectar she smelled in the hive during her first weeks of life (after emergence).

Scents are also used for communication. Bees produce information-rich odors, or pheromones, that organize the colony at every level. They play a key role in brood care, disease management, foraging, reproduction, defense, and more. (Read more on page 110.)

Taste

It should come as no surprise that honey bees have a sweet tooth. That is, they have sugar-sensing taste neurons that induce the kind of ecstasy some of us only get from chocolate. When a bee dips her tongue into a flower and tastes the sugary nectar inside, the neurons in her mouthparts fire intense pleasure signals that can last up to 10 seconds.

Her ability to taste sweetness isn't just about enjoyment, though. It's how she assesses nectar quality. Instead of taste buds, a bee has tiny, sensory hairs found mainly on her tongue, antennae, and forelegs. With them, she can detect different types of sugar (sucrose, fructose, and glucose) and their concentrations. This is important because not all nectar is the same—its components and sugar concentrations vary by flower species. Although most flowers generate nectar with 30 to 40 percent sugar, some provide as little as 3 percent, while others offer as high as 87 percent.

Honey bee nectar preferences vary among colonies and even among individual bees of the same colony. An individual's preferences might be influenced by colony needs, her assigned task, genetics, time of year, or weather. She uses her sense of taste to select the nectar that best suits her and her colony.

Honey bees aren't just sugar connoisseurs. They likely experience all five of the basic tastes: sweet, salty, sour, bitter, and umami. Their sense of taste helps them to choose the most profitable sources of food, water, and resins while avoiding harmful substances. Like us, they have a penchant for salt and umami flavors, but they aren't too keen on sour and bitter flavors. Foragers seek salty water for their own metabolic needs, but they also collect it to feed to their larvae. The claws on their feet are especially attuned to tasting salt. Water foragers prefer brackish ponds and can detect salt levels simply by dipping a toe. Umami flavors are an indication of amino acids, which, like salts, are essential to colony nutrition. Foragers prefer pollen that is rich in amino acids, and they may use umami taste receptors to help identify valuable pollen sources.

Touch

A honey bee has no sense of personal space. Inside the crowded hive, she touches another bee as often as once every two seconds. She greets her nestmates face first and pats them thoroughly with her antennae. In the dark of the hive, her antennae help her to recognize her sisters. Yet her touches are more than just a handshake—they are a means of communication. During nectar exchanges antennations are so vigorous, it looks like two siblings having a slap fight. In reality, they are signaling their readiness to give and receive nectar.

Some tactile forms of communication are even less subtle. One forager bee will headbutt another to warn her off a dangerous or unprofitable food source. This action along with the vibrations it delivers is called the stop signal (more on page 106). Touching each other may also be a way for the bees to determine colony size, a metric that influences colony decisions. For example, worker bees will not raise drones if their colony is too small to support them. Instead, they wait until their

Two hivemates cross antennae in a tactile exchange, but the information relayed is a mystery to us.

population reaches four thousand bees—but how do they know? One study found that colonies with an increased amount of physical contact between bees were triggered to build drone comb.

Although a honey bee has tactile sensors all over her body, they are more concentrated on the tips of the antennae, feet, and mouthparts. She uses these sensitive points to feel vital information. Her antennae, in particular, reach out like fingers to feel everything, even the surface she walks along. Researchers have found that bees can learn through touch. With their other senses obscured, they have been trained to recognize three-dimensional objects, tactile patterns, coarse and fine textures, and features like edges and grooves.

A bee's sense of touch is important at all life stages. In her larval state, she uses the texture of her cell walls to orient herself properly for emergence (with her head toward the opening). Later, when constructing new comb, special gravity-detecting hairs at the base of the neck help her to feel when the comb is plumb. As she shapes the paper-thin cells of the comb, she gauges the thickness and smoothness of the beeswax with her mouth and antennae. Once she starts foraging, sensory hairs on her antennae help her to feel how far she's traveled. On flowers, she can identify the species by reading the microscopic texture of the petals like braille.

Hearing

Honey bees buzz, hum, roar, beep, and even quack, but they don't hear the sounds they produce—they feel them. They "hear" by sensing the vibrations sound waves make as they move through substrates or disrupt particles in the air. Inside the hive, bees generate sound with their wings or thoracic muscles. When a bee buzzes, her nestmates feel her vibrations through the comb via their subgenual organs, special sound-perceiving organs found in many insect legs.

Comb, especially uncapped cells, is the perfect medium for vibratory messages. In natural beehives, the bees design their comb to amplify communication. Each piece of comb is attached to the cavity walls only enough to stabilize it, with the bottom edge left to hang freely. In contrast, most beekeepers encourage their bees to attach their comb to wooden frames on all four sides. Yet this practice muffles high frequencies, which may explain why bees so often thwart beekeepers by chewing the bottom edge loose.

A honey bee can also pick up airborne sound signals with her antennae, although not like we can. The Johnston's organ, located inside the elbow of the antennae, has hairs that are strung tight from end to end like guitar strings and can sense vibrations in the air when they are close to the source. These vibrations are interpreted by the brain as sound, making the Johnston's organ the closest thing a bee has to an ear.

We humans hear the different frequencies of bee vibrations as pitch. Some are lower than the lowest tuba notes, too low for human ears at 10 hertz. Others are on par with soprano singers, sounding higher than 1,000 hertz. Acoustic communication is thought to play a major role in the organization of the hive, but we are just beginning to understand it.

Magnetoreception

Another mysterious aspect of the honey bee's sensory world is her ability to detect a magnetic field. Earth, with its magnetized core, is looped by bands of magnetic force that are strongest at the poles. Like a living compass, a honey bee may be able to orient herself by sensing this invisible force. When a swarm of bees moves into a new hive cavity, if they are not confined by frames, they will choose to build their parallel combs in the same direction as did their parent colony. They remember the angle in relation to the earth's magnetic field, and, for unknown reasons, they replicate it in their new home.

The honey bee's method for magnetoreception is debated, but potential receptors have been found in the abdomen. Bands of iron-containing cells in a type of tissue called the fat body start to accumulate after a bee emerges from her cell. As she ages and consumes more iron-rich pollen, the iron granules in her abdomen increase in number and size. Interestingly, a bee takes her first orientation flight on her sixth day of adulthood, just before the greatest increase in iron, which occurs between days three and six. Her iron levels continue to climb as her orientation flights progress to foraging, but they eventually plateau on day nine. It may be that magnetic sense is one of many methods of navigation honey bees rely on for locating forage, but this is not known for certain.

Forecasting the Weather

Honey bees make excellent meteorologists. In fact we might be better off getting our weather forecasts from them than the local news. They can predict incoming storms by sensing barometric pressure changes. Not only can they anticipate bad weather, but they change their behavior to prepare for it. One study found that in the days before a big storm, foragers get ready in the same way we do—they stock up on extra supplies. Only instead of waiting in long lines at the grocery store, the foragers simply work longer hours to collect enough pollen and nectar to get through the rainy days to come.

I once witnessed bees' amazing weather-sensing abilities in my own apiary. I was standing in the garden a few summers ago when, suddenly, I noticed a change in the flight patterns of my hives. A huge cloud of forager bees appeared in the sky, but instead of flying back and forth like they normally would, they were all rushing home. Just a few minutes later, fat rain drops started to fall.

A forager bee takes a break while she waits out the rain.

CHAPTER 6

Buzz-words

The charming language of honey bees

Honey bee communication is as complex and beautiful as a symphony. Tens of thousands of bees must harmonize for their colony to thrive, and they do so simultaneously through dance, vibration, and pheromones. If we could cross the threshold of the beehive, I imagine it would be like stepping into a concert hall—an excited, humming place where each bee has a role to play in a grand, orchestral scheme.

The queen bee pauses in her work to dispense pheromones.

The Dance Language

At this very moment, honey bees are dancing in their hives. They move in the dark, swinging their abdomens and whirling on the comb. Honey bees frequently break into dance and have at least eight different types. They erupt in brief jigs and perform elaborate ballroom-like steps. Although other animals dance, bees are the only creatures with a symbolic dance language. This enchanting form of communication takes place day and night.

The Waggle Dance

Honey bees organize the collection of food and other resources with dance. When a successful forager returns to the hive, eager nestmates surround her. Their enthusiasm inspires her to perform the honey bees' most famous dance: the waggle dance. As she moves, she draws a treasure map on the comb. Her dance steps communicate the distance, direction, and quality of the flowers she's found so that other foragers may find them, too. Those who watch also learn the flowers' smell, which clings to the dancer's body, and even the taste of the nectar, because the dancer often hands out free samples. These sensory clues help the foragers pinpoint the exact location of the flowers once they leave the hive.

Like a pop star on a jutting stage, the dancing bee moves along a straight path, her audience rapt on both sides. She swings her abdomen laterally and buzzes her wings in short, rhythmic bursts. When she reaches the end of her path (her waggle run), she turns in a narrow semicircle to the place where she began and starts again. Each time she reaches the end of her waggle run, she turns in the opposite direction, giving her dance the shape of a figure eight.

The waggle run is the information-rich part of her dance. Its duration conveys distance. The longer she waggles down her path, the farther away the flowers are. Its course, which remains constant

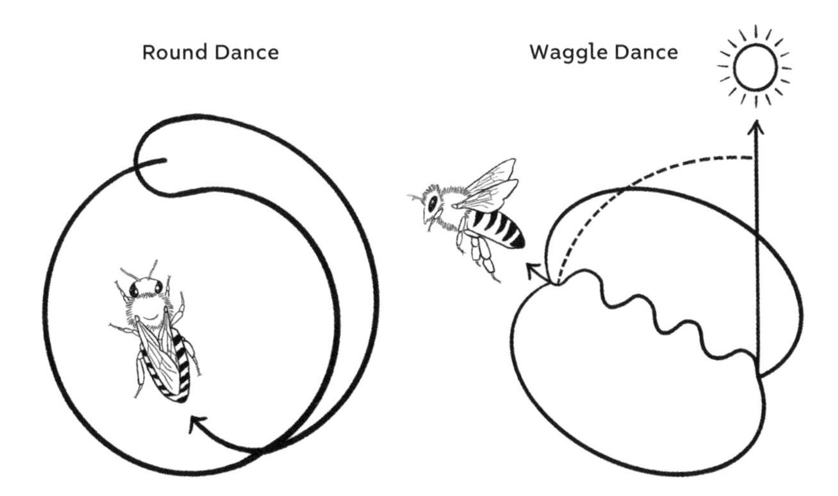

The round dance is used to direct bees to foraging sites that are close to the hive, while the waggle dance is used for sites that are farther away.

throughout the dance, communicates direction. The path she takes as she waggles up, down, or across the vertical combs represents an angle in relation to the sun. The watching bees somehow understand that the sun's position is always represented as directly up, like an ever-present disco ball. If a dancing bee moves upward during her waggle run, it means the forage is in the same direction as the sun. If she dances perpendicular to this imaginary line, across the combs, it means the forage is 90 degrees to the left or right of the sun. And so on.

The Dance Spectrum

The waggle dance is often described as one of three types of foraging dances, but we can also think of these dances on a single dance spectrum with two distinct dances at either end. On this spectrum the dance choreography changes from a single loop to a figure eight, depending on the flowers' proximity to the hive.

If flowers are farther than 100 meters, a forager bee uses the figure eight–shaped waggle dance. If the flowers are close to the hive, within 10 meters, she forgoes any waggling and uses a simple circular dance, aptly named the round dance, since foragers don't need as much information to find them. The round dance primarily serves as a prompt to leave the hive in search of the scent relayed by the dancer. The foragers don't fly in a designated direction but instead fly at random until they match the scent.

If the flowers are within 10 to 100 meters, the forager bee uses one of a series of transitional dances. These transitional dances are like the pages of a flipbook, with each page representing an incremental change in the flower's distance via a slight variation in dance shape. When the flowers are medium-close, the dance looks like the round dance, but as distance to the flowers grows, the pattern slowly morphs into the waggle dance.

Dance Spectrum

The dancing bee's choreography changes incrementally from the round dance to the waggle dance based on how far the foraging site is from the hive.

Dance Dialects

Scientists now believe that honey bees may have distinct waggle dance dialects. They have found that certain subspecies and breeds of honey bees communicate distance differently in their waggle runs. Therefore two bees from different genetic lineages may perform an identical dance with waggle runs that are equal length, but they are communicating different distances.

These dialects likely evolved as an adaptation to different terrains. Depending on the kind of biome the bees are in, they must travel varying distances to find food. Bees in a tropical setting, for example, on average do not have to travel as far as bees in a desert setting. If a desert honey bee used the same waggle dance length as a tropical bee to communicate distance, she would have to make impractically long waggle runs. This is why these kinds of colonies have adapted to communicate greater distances within shorter waggle runs.

Dance Classes

So how do new forager bees know how to communicate through dance? It turns out they learn it from experienced dancers. Like humans, young bees learn their language, in part, by observing and interacting with older family members. The honey bee dance language is complex and not easy to execute. If mistakes are made, those who watch the dance may never find the flowers they seek, which could mean less food for their colony.

A recent study removed all the older, dance-savvy bees from a colony to see how it would impact the dances. They found that bees who could not observe dances before performing their own made significantly more errors than the bees who had dance tutors. The same bees were tested later, and although their dances had improved, they continued to make mistakes with distance. This research supports the idea that distance dialects are tailored to the colony's environment and that this knowledge must be passed down through the generations of bees.

Dance Campaigns

A honey bee's dance is more than just directions—it's an advertisement meant to attract other foragers to a valuable food source. Like a sign spinner on a busy corner, she uses her dance to get the attention of passing bees. The better the source, the more exuberant her dance. So if the nectar has a high sugar content or the flowers are especially abundant, she dances like she's auditioning for a Broadway show. Her excitement draws more foragers to her dance, and when they return with their honey stomachs full, they perform their own excited dances, enticing even more foragers to the site.

As the bloom wanes, so do the number of dancing bees. Foragers switch to more productive sites and seek out new ones. In this way, dancers help the colony prioritize the best forage sites, which can shift throughout a single day as well as within a season. Dancing also directs foragers to collect based on colony needs. In spring, when pollen is critical for feeding nurse bees, dancers direct other foragers to pollen-rich flowers. On especially hot days when water is used to cool the hive, the dance floor is crowded with water foragers requesting backup. In the late summer when cracks must be sealed in preparation for winter, additional propolis foragers are enlisted through dance.

The dances are attended by foragers of all experience levels but are especially important for novice foragers, who rely on them to find their first food source 60 percent of the time. As a forager exits the hive, she has the option of scouting a new source on her own or stopping at the

A dancing bee is captured in a blur of motion as she waggles.

dance floor to pick up directions from another bee. Over her lifetime, she will embark on an average of one hundred foraging trips, and she will attend dances for about a quarter of those. Forage dances aren't just used for directing bees to new, unknown sources; they are also used to resume work at familiar sites. An experienced forager bee, for example, might briefly attend a dance in the morning just to confirm that the flowers she collected from the day before are still in bloom.

Beekeepers may have the privilege of watching their bees dance if they know where to look. Forage dances are performed on a designated comb near the hive entrance, called the dance floor. The dance floor is a crowded place and often features several dancers at once, each with her own audience of 1 to 10 bees. Although we enjoy the visual aspect of these dances, it's important to remember that the bees normally experience them in the dark. They don't watch the dances as we do; they sense them.

Dancing in the Dark

We don't fully understand the details of how the bees receive dance information, but there are likely several sensory cues involved. Nearby foragers who are drawn to dancing bees seem to collect dance information in different ways. Some gather around the edge of the dance floor with their antennae stretched toward the dancer. Others engage the dancer in a physical way, touching her with their heads and antennae. What sensory messages are they receiving?

When a bee dances, she makes her own music. On her waggle run, her abdomen sways in 15-hertz movements, 15 waggles per second. At the same time, her wings vibrate in rhythmic pulses, as if she were clapping. The rhythm of these high-frequency beats (265 hertz) may help other bees to follow and distinguish her dance in the noisy hive environment. The number of pulses could be how dance followers measure the length of the waggle run. Nearby foragers can detect dance sounds with their antennae. Interestingly, some antennal neurons associated with the ability to sense airborne sounds don't mature until the bees reach foraging age. Bees might also sense dance sounds as comb vibrations. Dancers who waggle on uncapped comb (best for transmitting vibrations) attract more recruits than those who dance on capped comb.

Foragers who attend a dancing bee closely follow her steps like dance partners, turning in tandem with her on the comb. They keep their heads perpendicular to the dancing bee's abdomen, maintaining contact with her via their antennae. Could they be following the pattern of her dance through touch? Are they measuring the angle of her waggle run with the orientation of their own bodies? Researchers continue to explore these questions.

Dance Scents

A dancing bee is perfumed with the scent of the flowers she has visited. When she returns to the hive, novice foragers memorize the floral scent and use it to pinpoint the location of the forage. The dancer is like a wise local giving a tourist directions to the nearest bakery. Only instead of telling them to look for the building with the striped awning over its entrance, she gives them a sniff of her baguette.

The floral scent is a detail that enhances imperfect directions. Because forage dances only provide an approximate location, foragers often fly in a zigzag pattern when they approach a foraging site for the first time. This technique increases their chances of encountering an odor plume that matches what they smelled during the dance. Once a forager has learned the location, she develops an efficient flight path based on memorized landmarks.

If an experienced forager recognizes a floral odor in the hive, she is motivated to visit the forage site where she encountered it before. This could be especially helpful in areas that have an abundance of a particular plant species. Many Southern California streets burst with purple in June when the jacaranda trees bloom. If a jacaranda forager returns to the hive carrying white pollen and a sweet, musky scent, those who are familiar with it can assume other known jacaranda sources are in bloom, too.

The Tremble Dance

Sometimes foraging conditions are *too* favorable. When a colony is faced with a sudden gold rush of nectar, the foragers don't lose their heads in pursuit of riches. Instead, they use a dance to restructure their workforce so they can take advantage of the boom.

When a nectar forager returns to the hive, she must unload her sweet cargo by transferring it to a receiver bee near the hive entrance. If she has trouble finding a bee to take her nectar, her foraging efforts come to a halt, but she doesn't remain idle. She performs a tremble dance to correct the task imbalance. She jitters across the comb, shaking her body from side to side as if she's had too much coffee. She walks a loopy, seemingly random path, yet her dance, which lasts an average of 30 minutes, has two important effects: It enlists more bees to receive and process the large influx of nectar, and it temporarily halts recruitment of new nectar foragers. In this way, the tremble dance keeps the bees efficient and helps them adapt when they come across a floral mother lode.

The Stop Signal

Foraging isn't all sunshine and nectar. Honey bees have predators, and many of them lie in wait in the flowers. When a forager bee narrowly escapes a spider attack, she returns to the hive with a warning. She charges onto the dance floor and delivers a forceful headbutt to any bee dancing for the dangerous site. Like a living fire alarm, she repeats a high-pitched chirp, vibrating 380 times a second, as she rams into the dancers. She often stays in the hive for a prolonged period and may deliver more than two hundred stop signals.

Although the dancers who receive the stop signal pause their dance in acknowledgment, they don't always stop dancing. This is especially true in small colonies. A recent study found that foragers in small colonies are more likely to ignore stop signals than those in large colonies. This is presumably because small colonies are in dire need of food and are willing to take risks to get it. The stop signal functions like a bad Yelp review: It may put off some diners, but those who are truly hungry will risk being murdered for a slice of pizza.

Grooming Invitation

A honey bee's body is designed to attract pollen, but it also collects debris and parasites. She addresses this by frequently brushing herself, licking her bristled legs and swiping them across her body like a cat. Her hive is also full of specialized groomer bees who proactively clean their nestmates. If a bee needs a deeper clean, she can request assistance by dancing.

She grips the comb with her legs spread wide and vibrates as she rocks her body from side to side. She often pauses between rocking sessions to clean herself in an energetic, exaggerated way, as if to say, "Hey! I could use a little help here!" while demonstrating that she cannot reach a hard-to-clean area. Grooming invitations typically last only 10 seconds, but they are highly successful in attracting groomer bees. A groomer often responds to a dance within 30 seconds. She climbs on top of the dancer and uses her mandibles to clean while the dancer extends her wings in accommodation.

By regularly cleaning nestmates, groomer bees play an important role in colony health, but, ironically, this particular job may increase the groomer bees' chances of contracting pathogens and parasites. Amazingly, a recent study found that groomer bees could have greater immunity protection than the average worker bee. Their blood (hemolymph) was more effective at clearing out harmful bacteria than the blood of other workers.

The Shaking Signal

Inside the hive, honey bees may motivate nestmates to work harder by pulling them into a spontaneous dance party. The motivational bee grips her sister with her front and middle legs and vibrates enthusiastically while shaking her abdomen up and down. Bees that are accosted in this merry way respond with increased bouts of activity. They move faster and accomplish more of whatever task they are engaged in. Like a hype man at a hip-hop show, the dancing bee roams the comb and uses this shaking signal to energize her sisters. It is believed that the shaking signal is a way to prepare the colony for periods of increased activity or a shift in tasks.

In support of this theory, the shaking signal happens more frequently during peak foraging season. Foragers returning from their first few trips to a new foraging site initially use the shaking signal. As they continue to visit the site, they gradually shift to the waggle dance. Perhaps this is a way to prepare the colony for the increase in nectar that a new foraging site will bring.

The shaking signal is also used on queens just before major events in their lives. For example, a queen may receive the shaking signal as she prepares to swarm, battle a rival queen, or mate. As drones approach maturity, they, too, receive the shaking signal. Drones who are signaled interact more with their sisters and enjoy increased grooming and feeding. Perhaps this pampering is a way to prepare them for puberty.

The shaking signal occurs hundreds of times per hour, but only 13 percent of bees perform it in their lifetime. Typically, they are older and associated with foraging, but the dance has also been observed in connection with other ages and tasks. Mysteriously, the number of shaking signals in a hive increases at night, but this might be because all the forager bees are inside the hive at this time. There is still much to learn about the shaking signal and its purpose. Even its name is debated. It's also called the dorsal-ventral abdominal vibration (DVAV), but maybe we will rename it the hype dance if more evidence in support of this theory comes to light.

The Buzz Run

When hanging from a tree, a honey bee swarm is extremely calm. Although the cluster contains thousands of bees all layered on top of one another, hardly any of them move. The surface of the swarm is as serene as a lake until the scout bees begin their buzz runs. These disruptive dancers plow through the placid cluster like they are trying to instigate a mosh pit. They run in a zigzag pattern, buzzing their wings in bursts and vibrating their bodies. They scramble over their sisters and disperse groups of inactive bees by charging into them.

They punctuate their dance further by launching into flight and zipping around the swarm before landing again to continue their buzz run. As a result, the surface of the swarm becomes visibly agitated and active like sloshing water. The intensity of the movement builds until the moment of flight when the cluster suddenly dissolves and reforms in the air as a massive, swirling cloud. The buzz-run dance appears to be an important preflight ritual that lets the bees know it's time to go!

Pheromones

The beehive is filled with purposeful perfumes. Many are chemical messages called pheromones that influence nearly every part of colony life. Pheromones are produced by every member of the hive: workers, drones, the queen, and even the larvae! Some pheromones are ephemeral. When the bees need to organize around a temporary circumstance, like an attack on the hive, they release these quick, pungent signals. After the bees have successfully defended themselves, the pheromone recedes. Other pheromones waft through the hive continuously and affect colony behavior for longer periods, like an olfactory set of house rules. These complex chemical signals allow the bees to communicate and organize their daily tasks.

Alarm Pheromone

The honey bee security system uses scent instead of sound to alert the colony to potential incoming threats. Guard bees at the entrance recruit defenders from within the hive with their version of a blaring siren: a banana-scented pheromone. It's produced by glands near the stingers and mandibles of mature worker bees, and it really does smell like bananas. When a guard sounds the alarm, she raises her abdomen, arching it menacingly with her stinger exposed as if brandishing a knife. She quickly spreads the pheromone to defender bees inside the hive by fanning her wings. Those who get the message fly from the hive ready to buzz, chase, and sting.

Defender bees also use the alarm pheromone to call for backup. When they deliver a sting, the scent works like a bullseye that marks their target's vulnerable spot and labels them as a threat. If the marked intruder doesn't leave the area immediately, more defenders arrive, and the response multiplies with each sting. These tactics are usually enough to ward off intruders for good, but should they return the next day, they

A group of helpful worker bees raise their abdomens, with Nasanov glands exposed, to guide new foragers home.

are met with a swift response. The pheromone marker left behind after a sting can last for several days, especially on fur or bee suits!

The Nasonov Pheromone

On spring afternoons, the beehive is like a chaotic airport. Soon-to-be-foragers add to the already busy traffic as they learn their surroundings through orientation flights. They are routed home by guide bees who fan a citrusy pheromone into the surrounding airspace. The guides line up along the entrance to the hive with their abdomens raised high, as if embodying the orange batons waved on a tarmac. Their strong lemongrass scent is produced by the Nasonov gland located at the tip of the abdomen.

The Nasanov pheromone works as an attractant, but, like our own language, its exact meaning shifts depending on the context. When a swarm changes location, the Nasanov pheromone is used to mark the spot where the bees will gather. Beekeepers who capture swarms know they've done it successfully when the bees raise their abdomens up and begin to fan on the hive box.

This scent is also used by scout bees to earmark promising forage locations and potential nesting sites. Inside the hive, it serves a similar purpose when a colony needs a new queen. Nurse bees release the citrusy pheromone near larvae they've selected for queen rearing. Not all of the marked larvae develop into queens, but those who are anointed with more of the lemony scent have a higher probability of becoming royalty.

The Queen's Bouquet

The queen has an almost gravitational influence over her colony. Her pheromones are an ever-present force that keep the colony together and organized. When she pauses on the comb, a retinue of bees encircles her. They reach toward her with their antennae, each striving for contact. Retinue bees feed and groom the queen, but they also help spread her pheromones throughout the hive. These pheromones let the colony know that the queen is healthy and productive; ensure that she is cared for; prevent workers from raising replacement queens; and stimulate worker bees to keep the hive clean, raise brood, and collect food. Without a queen, the colony falls into disarray and will eventually collapse.

The queen's pheromones are passed through the hive by touch. Many of them come from glands near her mandibles and are easily transferred to the nurse bees who feed and groom her, but they also linger on the comb. The bees in the queen's retinue are constantly changing. Some only attend her for 30 seconds before they are replaced by another. This high turnover rate gives the queen more opportunities to spread her scent.

It's often said that the queen's only job is to lay eggs, yet she takes frequent breaks from this task. She spends about half her time standing still on the comb. She even tucks her legs up underneath her body as if she were making room for more bees to crowd in and touch her. Could she be deliberately dispensing her pheromone bouquet? The surrounding bees frequently lick the queen and then groom themselves, which spreads her pheromones all over their bodies. They then become envoys of her royal scent. Over the next 15 to 30 minutes, they actively deliver the queen's pheromone message to other bees by feeding and touching them at an increased rate. Each bee who receives the message passes it on.

In the spring the queen's pheromones take on a new role. As the colony's population grows, the queen's pheromones become more difficult to disperse throughout the whole hive. The brood nest is so crowded that the bees on the periphery don't receive her message, and so they begin to raise new queens.

As the new queens are reared, the elder queen prepares to leave with a swarm. Thousands of worker bees take flight as they leave their mother colony. They form a massive, swirling cloud, and among them is their queen. As they fly, her pheromones act as a powerful attractant that keeps the swarm together with all the bees flying in the queen's orbit. As the bees journey to the new nest site, they take frequent rest stops. When the queen lands, she is like an anchor that pulls the swarm down out of the air to form a cluster around her. When they finally reach their new nesting site, the queen's pheromones motivate the colony to build comb and collect food for the eggs she will lay.

Brood Pheromones

Even though it's home to thousands of larvae, there's no crying in the honey bee nursery. Instead, the larvae attract the attention of their caregivers with pheromones. As a larva grows and her needs change, so do her pheromones. Her shifting chemical signals instruct the nurse bees on what to feed her, how warm she must be, and even when to cap her cell.

A larva's pheromones also have a surprisingly widespread effect in the hive. Not only do they influence the nurse bees, but they also alter the behavior of the foragers. Nurse bees excrete a special milklike food for young larvae, and in order to make it, they must gorge on pollen provided to them by the foragers who collect it. If nurse bees don't have access to pollen, they quickly lose the ability to feed the larvae in their care. In her first days of life, the larva emits a pheromone that encourages foragers to collect pollen. And foragers in colonies with young larvae collect twice as much pollen as those in colonies without.

Young brood also has a colony-wide effect that acts like a royal decree: Their pheromones suppress worker ovary development, which keeps them from interfering with the queen's role as sole egg-layer in the hive.

Footprint Pheromone

When a honey bee walks, she leaves behind an oily secretion produced by glands in her feet. This invisible trail, dubbed the "footprint pheromone," is attractive to other worker bees and is used in a variety of ways. It builds up on the surfaces inside the hive over time like layers of old paint. Could the accumulation of footprint pheromone explain why bees prefer used hive boxes to brand-new ones? With thousands of tiny feet crossing it each day, it's no surprise that there is an abundance of footprint pheromone at the hive's entrance. Unlike the Nasanov pheromone, which is fanned into the air to guide returning foragers, the footprint pheromone works almost like a welcome mat—it's applied to the surface in front of the entrance.

The footprint pheromone also guides foragers to floral sites. When a bee lands on a flower, she marks its petals, making it more attractive to other forager bees, especially those from her own colony. However, she does so indiscriminately. The petals are marked regardless of whether she was able to successfully forage from the flower.

Following page: Worker bees tend to their larval charges guided by pheromone signals.

CHAPTER 7

Drama Queens

Explosive sex scenes, death matches, and other tantalizing tales from the hive

The beehive is characterized as a hardworking utopia, but it's not without drama. Many bee behaviors could be fodder for supermarket tabloids. Perhaps unsurprisingly, the queen bee is often central to the colony's most rousing plots. Although her pheromones are what make the colony run smoothly, this peace is punctuated by dramatic life events. Even before she emerges from her cell, she could be murdered by a rival queen or imprisoned by the workers who raised her. This savage thread is woven into nearly every aspect of colony life, yet it is inextricable and even essential to the bees' survival.

A young queen emerges from her cell ready to face her rivals.

New Queens

The arrival of a new queen is like the introduction of a new character on a longstanding soap opera. She is a harbinger of change and upheaval that marks a transitional period for the colony, and she doesn't always survive. New queens are raised when the current queen suddenly dies, or when the colony is preparing to swarm, or when a queen is too old to lay eggs. A young queen's life can take several dramatic paths, but her fate depends on the circumstances of her making.

Emergency Queens

In sci-fi movies, when the alien queen is killed, the rest of her colony drops dead on the spot, but this is not the case with honey bees. The colony needs a queen to survive, but if she dies suddenly, the bees can usually replace her. To do so they must have a fertilized egg or a larva that is younger than three days old. Therefore the workers must act quickly, or their larvae will become too old to be converted into a quality queen. Within just 10 hours of losing their queen, the workers select several young larvae of varying ages and begin to build elongated queen cells around them. These special larvae would have become worker bees, but now they will be fed a strict diet of royal jelly, which will turn on different gene expressions and allow them to grow into queens.

We still do not know how the workers decide which larvae will become queens, but it seems they are not chosen at random. One study found that worker bees select larvae, at least partially, based on nutrition. In the study eggs that were heavier than others were more likely to be chosen for queen rearing. This makes sense, since other studies have found that queens raised from heavy eggs are larger and more fertile.

In an emergency queen-rearing situation, the colony raises many possible new queens, even though it needs only one. The process is not well understood, but the worker bees appear to be calling the shots. Not only do they select the larvae that will become potential queens, but they also decide which of these chosen larvae will survive to adulthood. Workers frequently destroy many of the queen cells they build before their occupants reach maturity. Researchers have found that they are more likely to cull cells raised from older brood versus those who were raised from eggs. It's probable that they raise multiple queens to ensure they end up with at least one that is of good quality. In other words, they don't like to put all their eggs in one basket.

Swarm Queens

Mature colonies replace their queens annually, through the process of swarming. Each year the queen ushers her colony through winter and into spring. She gradually increases her egg production, and as her colony's population reaches its height, she departs with a portion of them, a swarm, to establish a new colony.

The colony prepares for her exit by raising new queens from her eggs. In this case the workers do not select young larvae from the available brood. Instead, along the edges of the comb, they build small wax cups called queen cups. Their queen then deposits an egg in each one, and the workers begin the process of raising her replacement. One study found that the queen seems to know when she is laying an egg destined for royalty and that she may be able to control the quality of it. On average, eggs laid in queen cups are larger and heavier than the other eggs in the hive.

Over the next 16 days, these cups are extended into full-size queen cells to accommodate the growing larvae inside. The long, oval chambers are approximately the shape and size of unshelled peanuts. Although they hang obviously from the edges of the combs, they can be easily missed because worker bees often obscure them with their bodies. A colony may raise as many as 20 new queens for a swarming event,

but the queens will not begin to emerge until day 16, well after the mother queen has left with her swarm. Usually, the swarm takes flight when the first queen cells are capped, around day 8 of development. The mother queen leaves her colony with a plethora of immature daughter queens still growing in their cells, but only one can replace her.

Queen Detective Work

Experienced beekeepers are often able to discern why their colonies are raising new queens by the quantity and quality of the queen cells their bees make and by their position on the combs. Planned queen cells are well formed and large. When the bees make them to replace an aging queen, they are located on the face of the combs, and there are typically only one or two of them.

If the colony is raising queens for swarming purposes, its queen will have already left by the time the new queens emerge. Therefore the colony must make several queens for the best chance of having one succeed; these cells are classically constructed on the edges of the combs.

Emergency queen cells are made when the queen dies unexpectedly. In this case the workers also make a large quantity of replacement queens because their queen is no longer there to lay eggs, and they have only one chance to raise her replacement. Emergency queen cells often look droopy and small compared to planned queen cells and will have less texture on them.

A planned queen cell on the face of the comb contains a growing replacement queen.

The Queen's Choice

When a new queen emerges from her cell after the mother queen has swarmed, she faces a dilemma. She can compete against sister queens in a vicious battle to the death or rally a small, secondary swarm and flee. Both options come with considerable risk.

On the one hand, a rival queen could end her life with just one sting. On the other, the chance of survival with a small swarm is much lower than with an established colony. If she can best her sisters in combat, she will overwinter in a strong colony and leave with a large primary swarm the following spring. This is her best shot at long-term survival, although the path she takes is not necessarily a matter of free will. The worker bees can alter her fate. Although they do not participate directly in queen battles, they may influence a queen and her rivals toward their preferred outcome. Not only do they play favorites with individual queens, but if the colony population is large enough, the workers may collude to send out more swarms.

Queen Colosseum

And thus a new queen begins her adult life. She faces an unknown number of deadly opponents, and the worker bees that surround her may be tipping the scales in her favor or conspiring against her. The once-harmonious colony is now more like the Colosseum, a chaotic arena where queens must battle to the death among thousands of worker-bee spectators, and the loyalty of the crowd is fickle.

Once multiple queens are loose from their cells, the game begins, but it isn't resolved right away. Two or more queens may roam for hours within the same hive without fighting. Sometimes two competing

Two young queens, locked in deadly battle, attempt to sting each other.

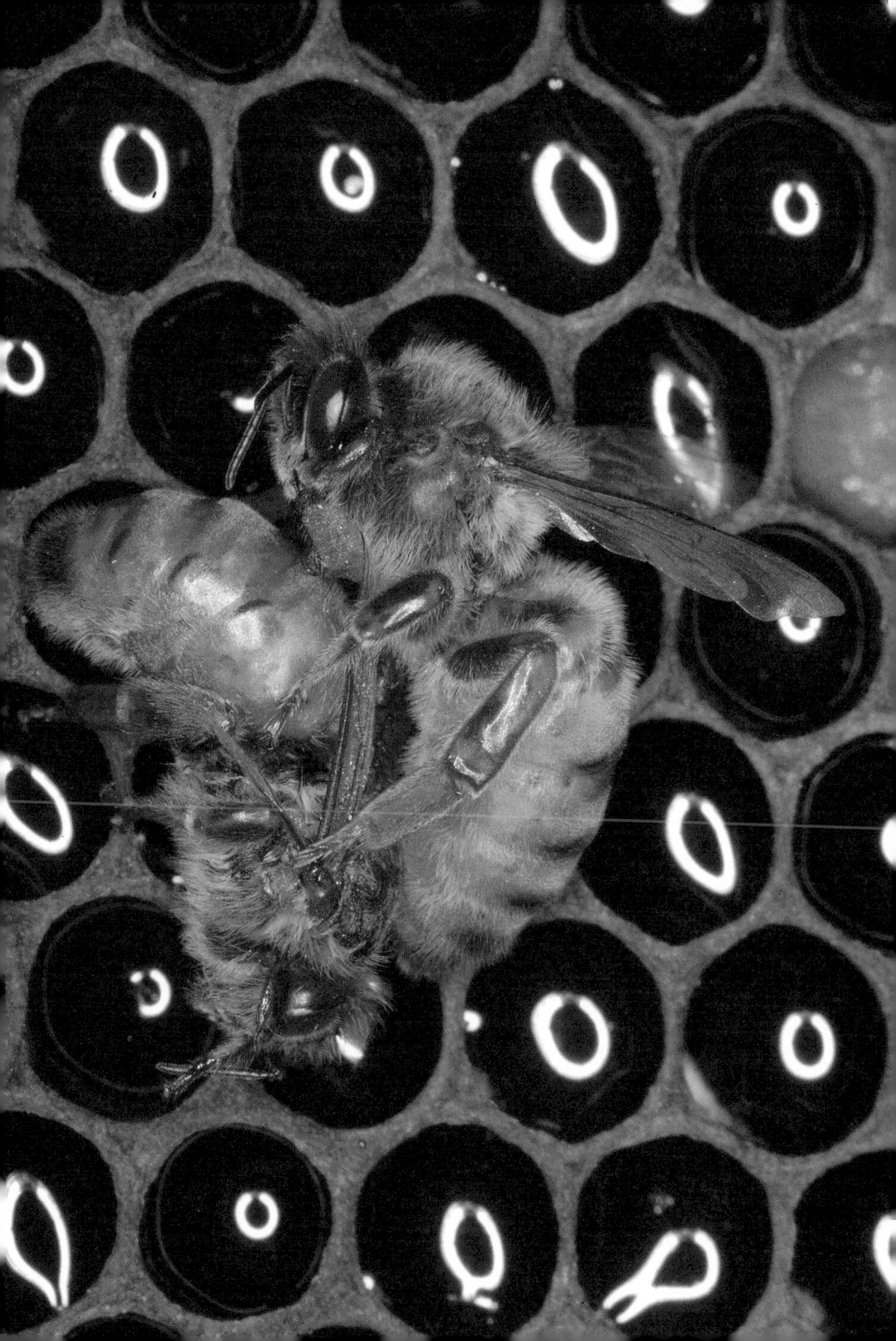

queens pass close by each other and never interact. At other times they meet face to face but part without aggression. But when one queen touches the other's abdomen, they immediately begin to fight. Researchers believe that there may be some kind of battle stimulus located between a queen's abdominal segments, something akin to a button the other queen must push; it appears to be activated only by physical contact.

The two queens become locked in a deadly wrestle. Each one attempts to gain leverage over the other. They bite, pull, and press with their legs as each desperately maneuvers her large abdomen into a stinging position. The struggle may last as long as 15 minutes or end in seconds. Once one of the queens has delivered a sting, her victim is doomed and often stops moving immediately, paralyzed by the venom. She will die within minutes, if not seconds.

Hide or Seek

Even if a queen emerges victorious from a queen battle, she may not be done yet. There could still be other queens for her to contend with. In this situation, a queen can employ several different strategies to survive. She may choose to hide inside a worker cell, often for several hours while (hopefully) the other queens eliminate one another. She may also decide to seek out her opponents before they emerge from their cells. New queens often patrol occupied queen cells, eagerly checking on them for the chance to do some damage. If she can, a queen will eliminate her competition without risk to herself. She simply chews a hole into the side of the cell and stings its helpless occupant to death.

Active queen cells are also patrolled by workers. Sometimes these workers remain neutral, but they may interfere. Workers sometimes protect the queen inside a cell by chasing away approaching queens. At other times they aid the attacking queen by helping her widen the hole she is chewing.

The Royal Treatment

Worker bees often interact with the queens who have emerged from their cells. They don't injure them, but they do vibrate, grab, and chase young queens around the combs. They do so even when there are no active queen cells nearby. The reason for this behavior is still unknown, but queens tend to move more quickly after they receive these attentions. Also, queens who last more than three hours often receive an increasing number of vibratory signals and tend to survive for longer periods, win more queen battles, and are more likely to become the colony's new queen.

Perhaps these workers are the bee-equivalent of cheerleaders. They boost their favorite queen's spirit and motivate her to succeed.

Sh*t Happens

A common battle strategy involves one queen spraying the other with her poop! Young queens produce a sweet-smelling rectal fluid that attracts worker bees. If a queen is sprayed by it, hundreds of workers gather around the unfortunate queen—immobilizing her for an average of 20 minutes. While she is trapped, her opponent can sting her more easily. Sometimes the rush of workers is so great, however, that the movement of both queens is restricted, and the battle comes to a temporary halt.

Either way, the strategy often pays off. A queen who sprays may gain a swift victory. At worse, she gets a respite from the battle. Researchers have found that in most cases, the queen who is sprayed ends up losing the fight.

The Queen's [Jail] Cell

Some worker bees use an extreme measure to protect a favored queen. Not only do they guard her cell from rivals, but they also prevent her from leaving it. When the imprisoned queen attempts to chew her way out of her cell, the workers may reseal it from the outside and even press their heads against the cap to barricade her. They leave only a small slit in the cell's tip for feeding their prisoner, whom they may keep contained for as long as one week. It's believed that the workers deliberately stagger the release of the queens so that each one may leave with a swarm instead of engaging in battle.

Tooting and Quacking

We all know that beehives buzz, but during queen duels they also quack! This is the curious cry of a queen bee who has been imprisoned in her cell. Her ducklike call is in response to an already mobile queen who makes her presence known with a toot. These sounds are collectively described as piping. They can be heard when there are several queens in the hive at once. It has been widely theorized by beekeepers that piping is some sort of battle cry, but researchers now believe that it's a way for the colony to prevent queen conflicts.

Once a queen is free from her cell, she often moves at a frantic pace but frequently pauses to toot. She lowers her body to the comb like she is getting ready to do a push-up and then emits a long toot followed by a series of shorter, more rhythmic toots. If there are any nearby queens held in their cells, they respond with quacking. They press their wings against the cell walls and produce short pulsating notes: *quack, quack, quack, quack*!

Although these sounds are delightfully audible to the human ear, the bees sense them as vibrations on the comb. This duet of tooting and quacking may be a colony-level communication system. It informs the workers that one queen is loose and gives them a sense of how many queens remain in their cells. Quacking queens are kept captive until the tooting queen leaves with a swarm. Each queen is released in this orderly way until there is only one left. When the last queen is released from her cell, she toots and gets no quacking in response. Her colony then knows that they have exhausted their pool of captive queens, and no more swarms will be sent out.

Victorious Queens

No matter how many battles she won or escaped, the victorious queen has survived against all odds. Yet she still has one final challenge before she can settle into her reign: She must mate. In some ways a queen is not truly a queen until she successfully mates. Before this act she is not fully developed. On the outside her characteristically large abdomen is still small, and it will not grow to its full size until after she mates. On the inside her sexual organs are still maturing, and the pheromones that attract worker bee attendants to her have not kicked in yet.

As she approaches sexual maturity, around day six of her adult life, the worker bees will begin to bite and chase her toward the entrance. Apparently they hope to annoy the queen so much that she will leave. This is because mating does not take place inside the hive—it takes place in midair! To mate successfully, the queen must abandon the safety of her hive and seek out male bees from other colonies. She prepares for this event in the same way that worker bees ready themselves for foraging: She makes several short, looping flights to learn her surroundings, and then, over the course of the next few days, she finally takes her mating flight.

The Mating Flight

Outside her hive, the queen is extremely vulnerable. Sadly, some queens never return from their mating flights. Research has found that as many as 6 to 26 percent of queens don't make it back to their hives. Though a queen may end up lost, stuck in a spider's web, or snatched out of the air by a hungry bird, she has no choice but to take the risk. Just as she has proven herself strong in battle, she must now prove herself fertile.

A victorious queen departs from
her hive on her mating flight.

The Drone Prepares

Every drone is on a quest to mate with a queen from a neighboring colony. He faces many obstacles but is endowed with useful gifts that may help him succeed. His extra sensitive antennae guide him as he flies, while his powerful wings sustain him in the air for long periods, and his massive eyes allow him to spot a queen moving at high speed. A drone doesn't reach sexual maturity until he is 6 to 12 days old, though, so he does not set out in search of queens right away.

A young drone typically spends his time in the brood nest. There, he is pampered, groomed, and fed by his sisters with increasing frequency as he continues to develop. He especially needs to eat pollen during this period because it is critical for sperm production. When he is near maturity, he begins to practice flying. Each afternoon in spring and summer, the drone takes flight. Just like the queen, he initially makes short, exploratory loops around his hive, but soon after, he embarks on longer mating expeditions.

The Meetup

Mating takes place at mysterious aerial sites, termed drone congregation areas, or DCAs. Like honey bee singles bars, DCAs attract thousands of eligible bachelors from nearby colonies who gather in hopes of meeting a queen. The number of drones at each site likely depends on how many colonies are established nearby. Most drones join DCAs within half a mile of their hive, but they can travel two to three miles. Research has found that a single DCA is typically populated with about 15 thousand drones, and they could be from as many as 240 different colonies.

These male bees fly high—as many as 115 feet (35 meters) above the ground. The space they occupy tends to be cone-shaped, like a giant party hat suspended in the air; its wide base may stretch 100 to 650 feet (198 meters) across.

The hopeful drone circles in this very specific area for many minutes, and then, as if barhopping, he moves to a neighboring DCA. To get there, he takes the same route his fellow drones have traveled before him. Like DCAs, these flyways are used year after year. Unless he is traveling between DCAs or returning home, the drone is reluctant to leave the designated mating space. He seems bound by some invisible perimeter and will not leave it, even if there is a queen just outside the bounds.

Inside the DCA, he faithfully circles until his energy begins to wane. The amount of time a drone can spend flying depends on how far he is from his hive. He typically spends about 30 minutes outside the hive before returning to rest and refuel. He takes only a 10- to 15-minute break before heading out again. Drones are most active in fine weather and are more likely to leave the hive at the warmest time of day. In most regions, this window is in the late afternoon. If weather permits, the average drone takes about six flights per day.

We still don't understand how DCAs are selected or how the drones find them. Somehow the drones return to these very specific locations year after year and use the same routes to travel to them. Some ongoing DCAs have records that date back 40 years, and one, in Selbourne, England, has apparently been active since 1792, first recorded by naturalist Reverend Gilbert White. This is especially curious since drones cannot live beyond a single mating season. Yet each spring the new generation manages to gather in these same locations without guidance from experienced drones. One study found that when drones were moved to a new location, they were able to find nearby DCAs within just 15 minutes, even though they were completely unfamiliar with the area.

Some researchers have found that DCAs share certain landscape features such as depressions in the horizon. DCAs have also been associated with places that have high levels of solar irradiance, which means places where more solar energy reaches Earth's surface. Pheromones may also play a role leading drones to DCAs, since sexually mature drones have pheromones that attract them to other drones.

The Act

The ready queen flies into the DCA at high speed, and even though she is surrounded by thousands of drones, she isn't easy to catch. As if playing tag, she moves this way and that, constantly changing the direction of her flight. Although the eager drones may get a whiff of her pheromones, they are unlikely to find her this way. The sheer number of beating drone wings disrupts her scent trail, and any drone following could be sent swirling off course. For that reason the drones rely mainly on their vision to find her. Research has shown that drones in a DCA are quick to react to any fast-moving object that enters the space, including butterflies and tossed stones.

The queen streaks across the sky like a shooting star, a comet tail of drones following close behind. The fastest drone must now maneuver carefully to get in the proper position for mating. This will all be over in a matter of seconds, but the steps are complex. First he accelerates his flight so that he is no longer behind the queen but just above her. He touches down on her abdomen, his head aligned with her middle, and grips her tightly with his legs. Somehow the queen continues to fly with the drone clasped to her, and he, too, keeps pace with his whirring wings.

Now the drone must curve his abdomen downward so that its tip is aligned with the queen's sting chamber. Then, with a sudden contraction of his abdomen, he pushes his penis out of his body and into the queen. This action is kind of like blowing air into a disposable glove. Up until this point, the drone's densely folded penis has been packed inside his abdomen, and it takes a rush of nearly all the blood in his body to inflate it. The total loss of fluid leaves him paralyzed and no

A successful drone mates with a queen midflight. His reward is death.

longer able to fly. He falls backward, and the queen sinks under his weight, losing some elevation.

The drone remains attached to the queen thanks to the shape of his penis, which resembles an abstract balloon animal whose shape fits into the sting chamber like a cork in a bottle. From there, the queen must finish the act by contracting her own muscles to complete the eversion of his penis and transfer the semen. As this happens, the drone falls farther back, and his penis detaches from his body. It remains lodged inside the queen, and the successful drone drops to the ground and dies, his quest complete.

The Sperm Bank

After the first drone falls away, the other drones are not deterred. They continue to line up behind the queen like flying dominoes. Each one removes the penis of the drone before and replaces it with his own so that the queen must leave a macabre trail of discarded body parts and bodies in her wake. Indeed she mates with an average of 12 drones, but some queens have mated with more than 40. This could happen on a single mating flight, or she may take a couple of different flights over the course of a few days. Despite the promiscuity of her early life, once the queen begins laying eggs, she will not mate again. Instead she stores the semen she has collected inside a special organ called the spermatheca and uses it throughout her lifetime.

Interestingly a queen's spermatheca holds about six million sperm, an amount she could easily acquire from a single drone, but she goes out of her way to mate with many more drones than seemingly necessary. The average queen may collect as many as one hundred million sperm, but she expels 90 percent of it. After she mates the sperm waits in her oviducts and then is pushed into the spermatheca through a complex compression of muscles. It travels in such a way that a little bit of sperm from each drone (5 to 10 percent) is saved and the rest is discarded. Once inside the spermatheca the sperm can live for as long as seven years.

Over time the sperm from each drone may find its way into the colony, but it is not represented equally. Some drones end up fathering more offspring than others. However, research suggests that specialty

A drone comet pursues a virgin queen.

worker bee jobs, like water collection or guarding, could be inherited from certain drones (patrilines). Could the number of offspring a drone fathers depend on how much the colony needs his genes? Either way it's clear that colonies with greater genetic diversity are better able to survive.

Bee Bromance

In most species, males compete to mate with the female. It's common for males to develop strategies to thwart their rivals, but this is not so with the honey bee. Drones do not try to stop one another from mating, even though they are technically in competition. This makes sense because it's in a drone's best interest for the queen to mate with a plethora of males. After all, a well-mated queen is better able to pass on the genes of each drone she has mated with.

Therefore a mated drone has developed a way to help the suitors that follow him. After he mates, his amputated penis (known as his mating sign) remains lodged in the queen and acts like a bullseye for the other drones. Studies have shown that, incredibly, drones have more success finding queens that have mating signs than those that do not.

The Mating Swarm

When the new queen exits her hive to mate, she is sometimes accompanied by a small group of worker bees. It's unclear whether they travel with her all the way to the DCA or if they are only a sendoff party. Perhaps it is a continuation of the "chasing toward the entrance" behavior that often occurs inside the hive with young queens.

Beekeepers who breed queens have also reported that the queen often comes back from her mating flight within a cloud of bees much bigger than the one she left with. A moderately populated mating nuc—a miniature beehive used in queen breeding—may end up loaded with bees after her return. It's likely that with so many new colonies close together, the neighboring foragers simply get caught up in the excitement (and pheromones) of a returning queen. As in a New Orleans wedding, bystanders are swept up into the fun and join the party.

Failed Drones

A drone's chances of successfully mating are not good. Instead of the Beach Boys' dream of two girls for every boy, honey bees offer one queen for every two thousand boys. This means that most males go unmated. Drones typically live three to five weeks in spring and summer, but some survive for several months. Older drones aren't fussed over by their sisters like young drones. They hang out in the periphery of the nest and tend to feed and groom themselves. Those who are still alive, after all, have failed to mate.

In most climates fall brings an end to the mating season, and the worker bees begin to kick out the drones. In what feels like a dramatic breakup scene, the drones are either turned away by the guards at the entrance or else physically dragged out of the hive. The unfortunate males are left to wander in the elements and will soon die of starvation or cold.

Although this seems like an unfair end, it is only done to preserve the colony. In many regions honey bees cannot afford to feed extra mouths through winter, and they especially don't need drones, since mating cannot take place during the cold months. Conversely, in some warm climates with year-round foraging available, it is common for colonies to keep drones all year. That said, drones are still sometimes booted from the hive even in these climates, typically during a sudden nectar dearth or an extended rainy period. It's therefore helpful to think of drones as a kind of luxury service that a colony is quick to give up when money gets tight.

Murder Ball

Once a queen has successfully mated, she remains in the safety of her hive until the following spring, when her colony will once again swarm. Now she is the one who departs with the large primary swarm, leaving her daughter queens to fight for the chance to succeed her. If

her swarm survives, she will repeat the process again next spring, and, in this way, she may live for several years. As she ages, however, her egg production and pheromones wane. This eventually prompts her colony to raise a replacement queen. Since the queen is still alive, the bees don't have to worry about their replacement queen failing, so they only need to raise one. If she does not develop fully they can raise another queen from their current queen's eggs.

Once the new queen has successfully mated, the worker bees may kill their elderly queen in a dramatic way: They cook her alive. The worker bees form a tight mass around their former queen and use their combined body temperature to heat her up. This ball of bees is gathered so densely that when handled, it feels like a prickly, hot stone. The workers make a strange, crackling hiss that sounds like broken power lines. They continue to heat the queen until she dies. This behavior is sometimes referred to as the "cuddle death," but beekeepers usually describe it as balling the queen.

Not much is understood about this behavior or why it is used instead of simply stinging the queen or expelling her from the hive like the drones. Sometimes, alternatively, an elder queen is permitted to live out her retirement alongside her replacement. She may even continue to lay eggs but at a reduced rate. It's likely that dual-queen colonies are more common than we realize, since most beekeepers stop looking for a queen once they spot one. This arrangement, however, is usually temporary. Beekeepers often report that the elder queen disappears after a few months. It is impossible to know whether she was killed by workers or by the replacement queen, or if she simply died of old age.

A drone expelled from the hive in fall is doomed to die.

Usurpation

Honey bees aren't typically thought of as cutthroat, but when it comes to survival, they can be downright villainous. If a colony swarms during summer months, that swarm is at a disadvantage. They don't have much time to prepare for winter. Some shrewd swarms make up for this fact by stealing an already established colony.

Often these swarms are smaller than the colonies they take over, so it's not clear how they manage it. But somehow they invade the larger colony, assassinate its queen, and replace her with their own. The bees in the invaded colony often seem to accept this transition. The swarm not only inherits the colony's comb, brood, and food stores but also its adult workforce. I suppose the losing worker bees know the old maxim: If you can't beat 'em, join 'em!

Two Queens to Rule Them All

In my bee rescue work, I am occasionally hired to relocate extremely large colonies and have found that they often have two laying queens. Initially I believed they were the result of a gentle supersedure, a mother and daughter queen living side by side, but then I noticed the queens were usually on opposite sides of the hive. I began to think maybe these colonies were so large that they required two queens to maintain their massiveness. Perhaps each queen oversees her own half of the comb and remains separate from the other, despite their shared body of worker bees.

My theory was strengthened when I realized that each time I encountered this scenario, the colony would experience a sudden and mysterious loss of worker bees soon after I transferred them to my beehive. This is likely because one queen went her separate way and departed with a large portion of the worker bees. But if one of the queens was old, with waning fertility, why would any portion of the colony leave with her? No, I suspect that these were two young queens, and in my effort to relocate the colony, I disturbed some delicate balance of power between them and prompted the colony to abandon its two-queen arrangement.

Still I could not think of how this colony with two young queens would arise in a natural way. That is until I received a call to relocate a pair of colonies living side by side. The twin colonies established themselves inside a single toolshed. They constructed their comb on opposite sides of the shed, and each one hung down from the ceiling like an upside-down pyramid. The two colonies had grown so massive that there was only a foot or two of space between them.

What would have happened if I hadn't been called to relocate them? Would the two separate colonies have merged into a single gigantic one, with a queen for each half? I now believe that this kind of merged colony is exactly what I had encountered in my previous two-queen removals.

CHAPTER 8

What Bees Don't Like

From bright lights to bananas, things that stress out honey bees and make them defensive

Honey bees only sting to defend their hive or themselves. Although some stings may seem unjust, the bee typically has a good reason for stinging. The problem is that most people have no understanding of bee behavior, so they can't comprehend why they have been stung. If a dog growls we know not to approach it, but when a bee buzzes few can discern its meaning. This seems a shame because bees are everywhere, and if we took the time to know them better, we might be spared the painful results of our misunderstandings.

Crab spiders ambush their pollinator prey by hiding in flowers.

Defensive Behavior

The first thing to know is that a honey bee is usually only defensive when she is near her hive. This is because she has a lot to protect, including precious honey and thousands of helpless babies. Plus, with so many hivemates nearby, she has ample backup to call upon if she detects a nearby threat. Every colony of bees is different, but most will ignore passersby if they aren't doing something that the bees perceive as a threat. Honey bees typically defend a 10- to 30-foot radius around their hive. If some unlucky soul draws their ire, the bees usually only chase them long enough to get them away from their hive.

When a bee is out foraging in the world, far from her colony, she is docile. She has nothing to defend except herself. A busy forager is only interested in her work, and she will not sting unless she is directly attacked. Rest assured you can closely observe bees on flowers and enjoy the hum of their labor with little risk to yourself, as long as you don't follow them home.

Painful Misunderstandings

Many people mistakenly believe that the only reason a bee might approach them is to sting. They don't realize that bees are curious and have a habit of investigating their surroundings. When a bee becomes interested in a person or object, she scans them from top to bottom by flying in a slow zigzag pattern. Her intention is not to intimidate but to learn. She may be searching for flowers or water or a nesting location. She could also be memorizing large landmarks to help her navigate.

Unfortunately most humans respond to this harmless bee behavior in the worst way—they swat. From the bee's perspective, this is a violent attack. A human hand is massive compared to the bee, and if it's attempting to knock her out of the air, who can blame her for feeling threatened? Sadly most people don't realize that their swatting response is what causes the bee to sting. When they are stung they take it as a confirmation that the bee intended to sting them all along, and their misconception is reinforced.

The best way to respond to a curious bee is to stand still and let her investigate. In some cases, however, a bee may be trying to communicate that she feels threatened. To do so she buzzes in a high-pitched tone and flies close to the face of her perceived threat in a rapid back-and-forth movement. Her quick, zippy flight pattern is like the annoyed tail flicking of a house cat. It's her way of saying, "Back off . . . or else." Those who do not heed her warning by leaving the area are often stung.

Some bee stings are purely accidental. A flying bee sometimes collides with a bicyclist or is stepped on near the pool when she is trying to get a drink. In these cases the bee is not malicious; her sting is an involuntary reflex due to physical harm. At other times a bee may become tangled in a person's hair or clothing and sting out of confusion and distress. It seems unfair to count these instances against her.

Genetic Behaviors

The honey bee has more than 20 subspecies (or breeds) that hail from different parts of their native lands in Africa, the Middle East, Asia, and Europe. Since they are all the same species each subspecies is capable of mating with the other, resulting in additional hybrid lines of bees. While breeds of dogs, horses, and chickens can look dramatically different from one another, honey bee subspecies look very similar. Some have distinct coloration or slight differences in size. And although their physical form doesn't change much, they can exhibit subtly different behavioral characteristics. Some breeds excel at honey production, while others are better adapted to cold climates. Some are favored by beekeepers for their docile nature, while others are notoriously defensive.

Apis mellifera scutellata is one such breed. One of 11 African honey bee subspecies, it was imported to Brazil in the 1950s to improve honey production in the tropics. There they mated with the existing populations of honey bees, creating several hybrids that quickly spread through South and Central America, eventually reaching the southern US in the 1990s. These hybrids are vigorous and disease resistant, but they can also be fiercely defensive of their nests and annoyingly (to beekeepers) quick to swarm.

Although many still refer to these hyper-defensive hybrids as "Africanized" or "killer" bees, these names are problematic and misleading. In general *scutellata* hybrids have the same likes and dislikes as other honey bees, but they are less tolerant than the docile Italian bees favored by most US beekeepers. When their nest is disturbed they may react up to 10 times faster, send out a greater number of defender bees, and pursue their perceived predator as far as a quarter mile.

Although this kind of extreme reaction can be dangerous, *scut* hybrids do not attack people or animals for no reason. Like all honey bees they only attack to defend their nests. And the intensity of their defensiveness varies by colony and region. In South America many beekeepers have embraced the ferocity of their *scut* hybrids and have adapted to their surly ways by wearing more protective clothing. In contrast Puerto Rico now enjoys a more amicable hybrid that has arisen through unclear means. Perhaps the bees were influenced by the laid-back island lifestyle.

What Stresses Bees Out

A number of stresses and disturbances, large and small, can rattle a hive. Some are obvious, like poisons and pesticides, but many others are not what most people would expect. Yet there are logical explanations for them all, if only we think like a bee.

Carbon Dioxide, Dark Colors, and Fur . . . Oh, My!

Honey bees recognize their predators in several ways, but most surprisingly by their breath! After all, hungry predators do tend to attack the hive mouth-first. Thus the bees have learned to associate carbon dioxide with an attack. Beekeepers like to joke that the bees don't tolerate bad breath, but the true reason a colony reacts to a forceful exhale is that it reminds them of an attacking bear, skunk, or badger.

For this same reason, honey bees dislike dark colors and furry textures. A forager bee looking for flowers will not attack someone on the street for wearing a black T-shirt, but if that person were near a hive, the bees might chase them off. Once a colony's guards have been roused, they tend to go after dark or furry targets first. They may also assail any red objects, since they see that color as black.

So if a blonde, a brunette, and a redhead walked into an apiary, it's safe to say the blonde would have more fun.

Bees Go Bananas

The honey bees' alarm pheromone smells a lot like bananas. So much so that beekeepers who wish to keep their bees calm may not want to eat bananas before inspecting their hives. If they do, the bees could mistake their banana breath for alarm pheromone and react defensively. This is because bananas and honey bee alarm pheromone both contain a compound called isoamyl acetate. The alarm pheromone has more than 40 different compounds, but isoamyl acetate is the main active component.

Once again, honey bees aren't going to attack someone in the lunch court just because they're eating a banana. Bees only react defensively toward this compound when it is near their hives. Even then, the scent must be strong enough for the bees to sense, and some colonies may be more sensitive to it than others. So it may be that some beekeepers can have bananas for breakfast, while others cannot.

Bright Lights

Honey bees don't have much of a nightlife. They spend their evenings inside the hive sleeping, making honey, or doing other household chores. Some may remain near the entrance and even just outside of it, especially when it's hot out, but they don't typically take flight unless they are disturbed by a predator or disoriented.

Since honey bees rely on the sun for navigation, they can easily become lost if they attempt to fly at night. Once outside the hive they are often drawn to bright lights. On occasion a single stray bee may end up flying in through a kitchen window, but if a high-powered security light is near a hive, it can draw hundreds of confused bees. They become agitated and fly at the light continuously until the point of exhaustion and death. Homeowners can avoid this issue by replacing their normal bulbs with yellow or red bulbs or by switching to motion-detector lights.

Zombie Predator!

One of the honey bee's weirdest foes is the zombie fly (*Apocephalus borealis*). This parasitic fly resembles a tiny fruit fly, but instead of laying its eggs in fruit, the mother zombie fly lays her eggs inside a living honey bee. The infected "zombee" then becomes disoriented and leaves her hive, often flying off into the night and ending up circling a bright security light. After she dies as many as 13 zombie flies can crawl out of her body.

Many beekeepers let the grass grow tall around their hives rather than risk the wrath of the bees by mowing too close.

Yard Work

Honey bee colonies are always on the lookout for threats to their hive, so it makes sense that they dislike it when yard work is happening nearby. Imagine peeking out of the hive to see a massive, vibrating machine charging toward your home in a cloud of gasoline fumes, shredding grass along the way. It's easy now to see why the guards would object to an approaching lawnmower. Tree trimming, too, may appear destructive and threatening to a nearby hive. Even digging or raking could set off a touchy colony. After all many honey bee predators are disruptive to the environment in similar ways, and honey bees have adapted to be wary of these behaviors. Beekeepers may want to don their suits when performing these activities near the apiary, and homeowners should take care to look out for wild hives in the vicinity before they start any large projects.

Queenlessness

Honey bees need their queen to survive, but if something goes amiss and they end up without one, they will continue to live for many weeks in a state of distress. Beekeepers may find that a colony that was previously docile is now defensive and difficult to work with. The doomed bees may be quicker to sting or else run around the combs nervously. They can make a desperate sound: a high-pitched whine layered over a low, disquieting roar. Those who are in tune with their bees can often sense that something is wrong by their behavior alone, but if they aren't yet sure, confirmation often comes after a thorough inspection reveals that there are no eggs or larvae in the hive.

Dearth

Honey bees get hangry. At the end of summer, when the flowers dry up and the bountiful flow of nectar slows, the bees must survive what beekeepers call a nectar dearth, and they aren't happy about it. This period

Hive and Prejudice

Every so often I meet someone who insists that bees simply don't like them. I usually quiz these people about their habits, looking for logical explanations for the bees' supposed prejudice. Does the person swat? Wear strong scents? Eat bananas? No, they say, and I have to let the subject go, but I never truly believed their claims. That is until a few years ago when I worked with the local zoo to remove a large wild colony.

The bees had built their hive just above a doorway in an off-exhibit area and had been there for more than a year. Their large, three-foot combs hung down, completely exposed except for where they were covered with bees. Amazingly they hadn't bothered anyone—with the exception of one employee. I was told that other people could come and go through the door without incident, but when this particular person approached the hive the bees became agitated and would chase her away, even though she had never done anything to disturb them.

I looked up at the bees who hung placidly just over our heads and didn't believe a word of the story. The colony was incredibly calm, despite half a dozen of us coming and going through the door with ladders and other equipment in preparation for their removal. They remained docile even after I began transferring their comb to my wooden hive box.

Then, halfway through the process, an employee came by with a roll of duct tape that I had requested over the radio, and the colony's entire demeanor changed. The surface of the cluster seemed to ripple with agitation, guards took flight, and the low hum of the hive shifted to a high-pitched buzz. The newcomer dropped the duct tape and took off running. I turned in shock to my team of helpers with the unasked question all over my face. Yes, that was the aforementioned person, the one the bees simply *didn't like*, and I never found any logical explanation for their prejudice against her.

is universally dreaded by beekeepers because their previously docile colonies become cranky, defensive, and unpleasant to work with.

Non-beekeepers, too, can be affected during these periods because the hungry bees descend on any sweet substance they can find. They mob trash cans in search of soda or juice and will completely take over hummingbird feeders. While some bee lovers may be tempted to put out extra sugar-water feeders for bees during this time, the best way to help is to plant late-blooming trees and shrubs. Although sugar water can help bees survive, it doesn't compare to the nutrition they get from real nectar. It can even create such a frenzy that bees begin to fight over it—to the death.

Robbing

Honey bees are usually good neighbors, but when flowers become scarce and the nectar dearth sets in, they can turn on each other. Some elder foragers take on the risky role of robber, and instead of seeking out nectar from hard-to-find flowers, they sneak into a nearby colony and steal from its stores. If a robber comes upon a small, vulnerable colony, she may return to her own hive and recruit even more bees to rob it.

The situation can escalate quickly as robber bees overwhelm the weak colony. The frenzied cloud of invading bees attacks and kills the opposing bees until there is no more resistance. The robbers take all that can be carried, leaving behind a pile of dead bees and chewed, empty comb. Beekeepers can prevent their colonies from robbing each other by making sure they are all of equal strength and by reducing hive entrances ahead of any nectar dearth so that there is a smaller area to guard.

Robber bees empty the honey cells of a weak colony.

Magic in a Jar

In spring and summer, when flowers are abundant, beekeepers give their colonies extra hive space in hopes that they will fill it with honeycomb. If all goes according to plan, the bees make more honey than they need to survive the winter, and the beekeeper can harvest a portion of it for themselves. Some beekeepers use the phrase "robbing honey," which is perhaps a more apt description of the practice.

I can't pretend that the bees are happy to give up their precious honey stores—it's not in their nature. I'll also admit that honey is not the reason I keep bees, and harvesting it is one of my least favorite beekeeping activities because it is sticky, time-consuming work. So why do I take honey from my bees? I personally think of my relationship with bees as a partnership, and, as in any collaboration, there are compromises. I tolerate their stings, and they tolerate my honey thieving. After all the bees are not imprisoned in their hives and are free to abandon them at any time.

To me a honey crop is like a profit-sharing check. I give the bees a place to live and do everything in my power to keep them healthy. During poor production years I leave what little honey there is for the bees, but if our combined efforts yield a surplus of honey, we share the spoils. The honey is just a bonus to me; I don't rely on it to keep my bee business afloat. Instead I focus on education and bee rescue work, but I find that honey is a valuable educational tool. It links people to what they have learned and motivates them to make positive changes to help the bees, which is why I only sell the honey that I harvest during my educational events. I want it to be truly appreciated for what it is—a little bit of bee magic in a jar.

Parasites and Disease

Honey bees live in close quarters. Like sailors, they are meticulous about keeping their small space clean, but they are still susceptible to parasites and disease, especially when they are stressed. Although a colony can sometimes recover on its own, a colony that is weakened by parasites and disease may grow overwhelmed and eventually die. Dying colonies are often affected by multiple stressors, such as mites, disease, and poor forage. The synergistic nature of colony failure often makes it difficult to discern what first weakened the bees; it's only evident that their ship is sinking.

Several parasites plague the honey bee, but none more so than the varroa mite. This troublesome mite can be found in every colony and throughout most of the world. Although its red, oval body is large enough to see, it's typically hidden from view because it burrows between the honey bee's body segments. Not only does it weaken the bee by feeding on her fat body tissue, but it also infects her with viruses that can quickly spread through the colony.

The trouble begins when a female varroa mite sneaks into the honey bee brood cell just before it's capped. After the cell is covered it becomes the perfect environment for the mite to lay her eggs. She lays both male and female eggs, and the resulting mites mate inside the sealed cell. When the bee emerges as an adult, so do the pregnant female mites that developed alongside her. Two weeks later these newly minted mites enter more brood cells and lay even more eggs. In this way the varroa mite population can quickly explode, especially when a colony has lots of active brood comb.

In nature some colonies can survive infestations, but it's not totally clear how. They may have stronger genetics, groom each other more often, maintain smaller brood nests, and swarm frequently. Swarming creates periods in which there is no brood in the colony. This lowers mite levels because the mites are unable to reproduce without developing bees. Unfortunately many beekeepers struggle with varroa mites in their colonies. The mites are often cited as the number one reason for colony loss, and there is much debate over the best way to manage them. Some beekeepers use chemical treatments or forced brood breaks, while others rely on strong genetics in isolated settings, but most agree that multiple strategies should be used.

Honey bee disease can spread via varroa mites or between colonies or from drone to queen while mating. Once inside the hive viruses spread easily between nestmates. One of the most common viruses, deformed wing virus, is transmitted by varroa mites. It is easily recognized because, as the name suggests, it leaves bees with shriveled, unusable wings. Unfortunately it often appears when varroa mite levels are high, and it marks the colony's imminent decline.

Although some honey bee diseases affect adult bees, most attack the brood. In a healthy colony the queen lays her eggs in the central part of the comb. She systematically deposits one egg per cell, creating a large grouping of brood that is close in age. The cells mature to the capped pupae stage, creating a solid patch of capped brood cells. If disease strikes, however, the adult bees do their best to control the outbreak by removing the infected brood. This practice creates holes in the patch of healthy brood, resulting in an uneven or spotty brood pattern where some cells are capped and others are not.

A spotty brood pattern is one of the first signs of an ailing colony and may be observed when the bees are struggling with a viral, bacterial, or fungal illness. As the disease progresses, the colony's population shrinks and no longer has the workforce to remove the infected brood. At this advanced stage of infection, the brood is visibly sick. Depending on the disease the brood dies in different ways and at different ages.

A varroa mite clings to the underside of a worker bee.

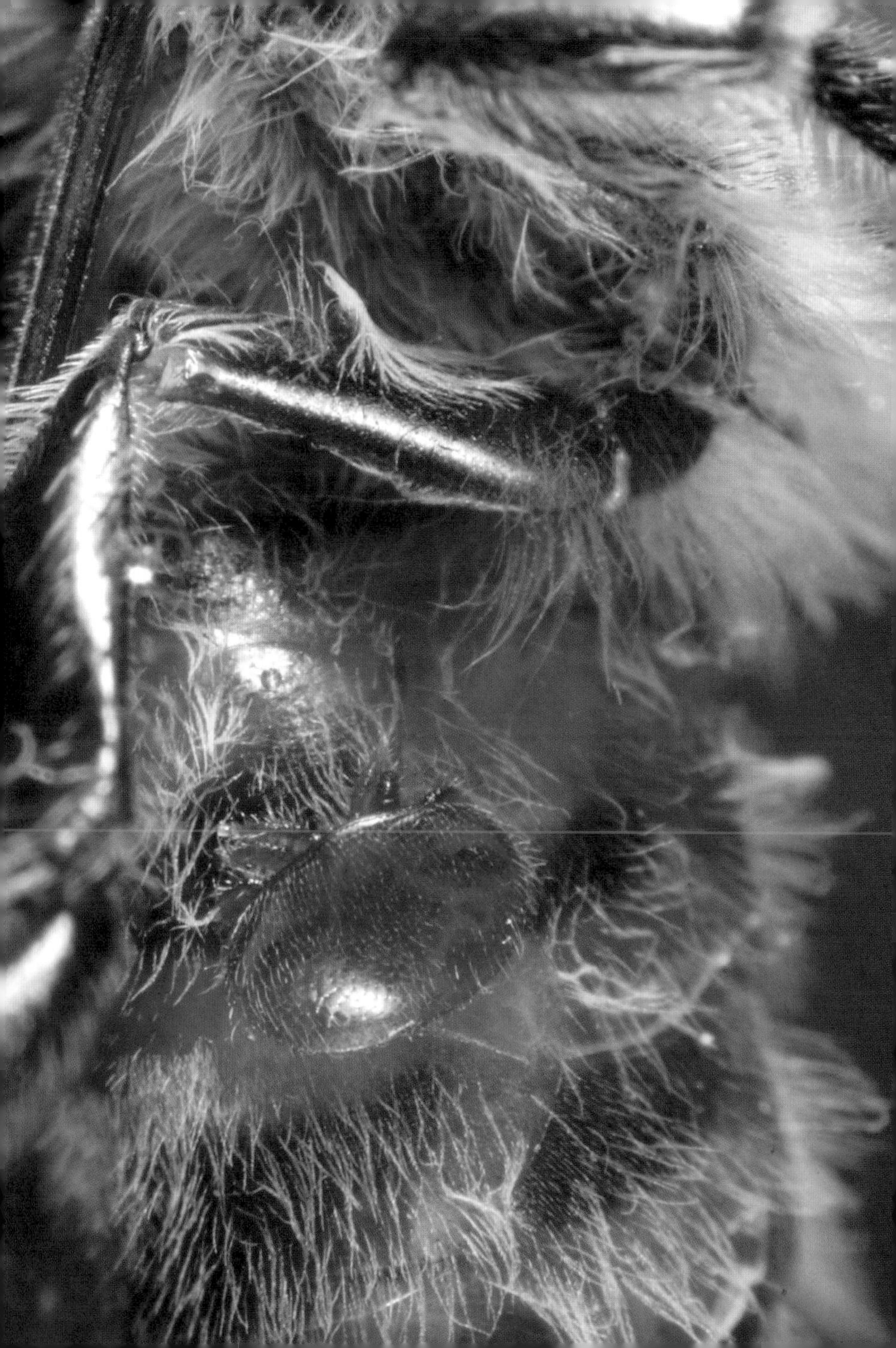

For example a larva infected with European foulbrood becomes twisted and yellow and dies before it has the chance to pupate, while American foulbrood kills off brood during its pupal phase, turning a would-be bee into stringy, brown goo.

Pesticides

Tragically one of honey bees' greatest strengths is also what makes them vulnerable to pesticides: their ability to forage. Because they forage on a large variety of flowers and travel miles around their hive to do so, their chances of encountering a plant that has been treated with a pesticide are extremely high. Pesticide contamination is widespread, especially in the US, where one billion pounds of pesticides are used every year. The US has also failed to ban many of the pesticides that other large agricultural nations have deemed unsafe for bees. A recent US geological survey found that more than 90 percent of waterways contain five or more pesticides. This is true in both urban and agricultural settings, but insecticide levels were highest in urban water samples.

While agricultural pesticide application is at least somewhat regulated, pesticides for the average US consumer are not. Anyone can purchase a bottle of insecticide over the counter and use it improperly in a way that could easily harm bees. Some homeowners may not even realize they are applying a pesticide because many are marketed as "all-in-one" plant care products that also contain fertilizer. Even those who never use pesticides in their gardens could unwittingly contaminate their pollinator havens through nursery purchases that may come pre-treated with pesticides.

Pesticides are typically applied to plants as a spray or seed coating. Honey bees are exposed to them when they forage on the pollen and nectar of treated plants, while collecting contaminated water, or, if toxic dusts or liquids are suspended in the air, as they fly. If a forager bee encounters a high concentration of pesticide while she works and manages to make it back to her colony with the poisoned load, the results are catastrophic. At the hive's entrance she passes the contaminated nectar to her nestmates, who, in turn, pass it to others. This can wipe out the entire colony with shocking speed.

One study tracked the nectar from just six forager bees and found that it was shared with 62 percent of all foragers in the hive and about a fifth of the house bee population in three and a half hours. After the poisoned nectar is passed through the hive, the once-booming colony is transformed into a graveyard. A thick bed, made up of thousands of dead and dying bees, stretches out in front of the entrance. Those who are still alive exhibit the telltale signs of acute pesticide poisoning. Unable to fly, they desperately skip and spin in circles. Eventually they flip onto their backs and twitch uncontrollably until they succumb. This scene is a beekeeper's worst nightmare, but unfortunately it can happen anytime a neighbor liberally applies a pesticide to a flowering plant.

Sometimes the effects of pesticide exposure are immediate and obvious, but most of the time they are subtle and invisible. These sublethal effects don't kill colonies outright, but they make it more difficult for them to survive. This is particularly true of neonicotinoid insecticides, whose sublethal effects impact honey bee cognition, metabolism, immunity, fertility, queen survival, and more. These harmful side effects often result in a slow death spread out over several months. When the colony finally collapses it often looks like the bees died from a virus or queen loss, but these are just symptoms of pesticide poisoning.

Unlike other pesticides, which are applied as a topical layer over the foliage, neonicotinoids are absorbed into the plant tissue itself. Once they are inside the plant, every part of it becomes toxic, including its leaves, pollen, and nectar. Even after a single treatment, the pesticide can persist inside the plant for months or even years. It can also accumulate in the soil and water of treated areas and spread to neighboring plants that were never treated.

Unfortunately neonicotinoids are the most-used pesticides in the world. Many consider this class of pesticide to be the driving force behind decreased survival in not only honey bees but other species as well. Recent surveys have revealed that insects are declining at alarming rates, along with many of the species that feed on them, such as birds, bats, and frogs.

Pollution

When we think of animals harmed by pollution, honey bees don't usually come to mind. New research has uncovered that honey bees are negatively impacted by air, water, and plastic pollution. When bees fly through the air, they are exposed to the particles within it. Their hairy bodies are designed to collect pollen, but they also gather debris like dust and dirt. In heavily polluted areas bee bodies can become contaminated with heavy metal particles and microplastics suspended in the air.

Some of these toxins may shorten a honey bee's lifespan and disrupt her ability to forage. Traffic-derived air pollution may also reduce a honey bee's ability to sniff out flowers and remember their locations. In one study, bees exposed to exhaust visited significantly fewer flowers than those who were not. Alarmingly air pollution can also impact the plants that bees rely on, disrupting their blooming cycles or resulting in fewer flowers overall.

The Popeye Effect

Honey bees' ability to survive becomes superpowered if they have plenty of food. When flowers are abundant, a honey bee colony is like Popeye with a can of spinach. A once-middling colony can transform into a buzzing force of nature. Its population explodes, granting it a generous workforce capable of combating predators, disease, and parasites. Not only that, but the availability of different flower types gives the bees access to a variety of nectars and pollens, some of which may help them overcome specific pathogens or parasites. A recent study found that sunflowers may protect honey bees from varroa mites. Another found that high-quality pollen can help bees metabolize pesticides and withstand their harmful effects.

Champions of the bees may take heart in these findings. While we cannot reverse everything that causes bees harm, we can at least improve their chance of survival by providing them with better nutrition in our gardens.

CHAPTER 9

Universal Bee

The honey bee's characteristics and why we admire them

Honey bees inspire devotion. They have captivated us for thousands of years, and still we study them. Our ancient reverence for honey bees is evidenced in Mesolithic rock paintings and Egyptian sarcophagi and is even minted on the coins of the ancient Greeks. Indeed many early societies were not only fascinated by honey bees—they believed they were sacred. Although they are no longer central to today's religions, bees continue to be popular in modern songs, poetry, and art.

What is it about the honey bee that we love so much?

Sunflowers are especially popular with honey bees because each flower head contains thousands of tiny nectar- and pollen-producing florets.

Economy

Life as a honey bee is expensive, metabolically speaking. On an average day a colony engages in a number of energy-intensive activities, often simultaneously, and all are fueled by precious honey. As a result honey bees are excellent at conserving their resources and are incredibly careful about how they spend them. The economical nature of honey bees is not only admirable, it's inspiring.

Foraging Isn't Free

In the beehive it takes honey to make honey. The colony's honey-making operation is a massive enterprise, but somehow the bees manage to make a profit. Their efforts begin with careful foraging.

Nectar is a difficult resource to collect because each flower offers only a tiny amount. A forager must visit hundreds or even thousands of flowers to gather a load, but she is careful not to spend more energy flying than the nectar is worth. Her energy expenditure depends on a number of factors, like how far the flowers are from the hive, the abundance of flowers available, the spacing between the flowers, and how well matched her body is to the flower type.

Certain foragers like certain flowers over others (see page 48). Unsurprisingly these preferences are driven by what is most energy efficient. Foragers minimize the amount of flying they have to do by visiting large clusters of tightly grouped flowers with inviting shapes that are near their hive. These kinds of floral resources don't just make for a quick round-trip flight; they also allow foragers to climb from flower to flower rather than fly, and flowers with open petals improve the speed of nectar collection. To sum it up, if a honey bee were grocery shopping, she would buy the bag of already shelled pistachios.

Energy-Efficient Hives

Nectar collection is only the first step in honey production. Once the foragers have gathered a load, they hand it off to house bees for the intensive process of ripening. The bees may spend more than 50 percent of the energy of each nectar load on its evaporation. Therefore, the foragers must appraise the value of their nectar load against its entire labor cost—which is why they also consider the energy efficiency of their own nesting cavity.

The required temperature of a colony's brood nest is often warmer than the outside air: between 90 and 97°F (32 to 36°C). So, when selecting a nesting cavity, scouts look for thick-walled hollows with good insulation to hold in the heat. These sought-after nesting sites aren't always available, but they can improve a colony's spring honey production. A thermally stable nesting cavity means that fewer bees have to expend their energy on warming or cooling the brood nest. Therefore the colony eats less honey and frees up its workforce for other activities, such as foraging. Plus a well-insulated hive is naturally warmer, even outside the brood nest where honey is stored. This helps honey to ripen faster and with less effort than a less insulated hive, because warm air holds more moisture than cold air.

A recent study found that when foragers have a thermally stable nest, they will travel farther from the hive to collect nectar. This suggests that foragers know they can afford the cost of a longer flight because of the energy saved on processing nectar back at the nest. Beekeepers should take note of these findings, since most hive designs have thinner walls than what is available in nature. As beekeepers, we have prioritized the ease of lifting hive boxes over the well-being of our bees.

The Cost of Comb

To store its honey the colony must construct comb. To do so the bees need to generate a tremendous amount of beeswax. Young bees excrete liquid wax from glands in their abdomens. The wax cools in thin, wafer-like flakes. Approximately one billion of these wax flakes are needed for the colony to construct its comb, but it's not a cheap building material. For every pound of wax, it's estimated that the bees must consume 8 pounds of honey. And the bees use it so wisely that when a pound of wax is turned into comb, it can hold up to 22 pounds of honey!

The bees achieve this efficiency through their use of the hexagon. Unlike circles, neighboring hexagons fit together perfectly and share walls,

Wax Recyclers

Given the value of beeswax, it should be no surprise that honey bees reuse it! Inside the hive worker bees sometimes remodel their comb. As if peeling an orange, they use their mandibles to bite off long, curling strips of wax from existing structures and repurpose it where needed. Since active wax glands are not required for this work, bees of any age may do this activity. Is this a way for older bees to help with comb construction when new comb is needed more urgently, or is it just for quick repairs?

I have also observed forager bees collecting beeswax from my solar wax melter, which melts the beeswax out of old comb using the sun's heat. If the bees get the chance, they will chew off bits of warm wax and pack it onto their back legs like pollen to carry back to the hive. Inside I have found newly constructed combs made of marbled brown and white wax, which I suspect is a sign that it was built from a mixture of new and recycled wax. As interesting as this is to observe, beekeepers should be careful not to let their bees forage on old wax because it can transmit a disease called American foulbrood.

Festooning worker bees work together to build new comb.

which means the bees can use less building material without sacrificing storage space. Human mathematicians have only recently proven what honey bees have known all along: that the hexagon is the most efficient shape. The hexagonal comb structure allows the bees to store the maximum amount of honey while using the least amount of wax to construct it.

The shared walls are also the secret to the honeycomb's strength. When hexagons come together, they form a series of three-sided joints that balance each other and create an incredibly stable structure. Given its strength and efficiency, it's no wonder that honeycomb has inspired human engineering. The hexagonal pattern can be found in structures the world over, including airplanes, products of 3-D printing, and the ancient architecture of the Pantheon.

Frugal Fertility

Honey bee economy is not limited to honey production and nest structure—it also takes place inside the body of the queen bee. The queen only mates at the beginning of her life, so she has a limited amount of semen from which to fertilize her eggs. If she mates well, she begins her life with five to seven million sperm stored inside her. This sounds like a lot, but she has to make this stored sperm last her entire lifetime, up to eight years. And in spring, she is expected to lay as many as three thousand eggs a day. If her fertility wanes, the worker bees will cut her life short and replace her with a new queen. It's therefore critical that she use her sperm frugally, and she does: Researchers found that the queen typically uses just two sperm per egg fertilization!

Balance

The beehive is filled with examples of near-perfect balance. From the structure of their comb to the flexibility of the job system, the bees seem to have it all figured out. Yet some examples are so elegant they deserve special recognition.

Hot and Cold Destiny

In most climates honey bees spend their summers foraging and their winters cloistered inside, eating honey and working hard to stay warm. Despite the outside temperature, the bees strive to keep their brood nest between 93 and 97°F (34 to 36°C). Incredibly, one study found that the temperature of each individual pupa can vary and that it influences the developing bee's future career. Capped brood raised at 90°F (32°C) grow up to be poor foragers, while capped brood raised at 97°F (36°C) become excellent learners and dancers. The researchers suggested that this may be by design to keep the colony's workforce in balance.

In summer the brood nest's temperature is at its highest, and so is the colony's need for excellent foragers. Then, during winter months, the brood nest cools, and so does the colony's need for foragers since there are few (or no) flowers. This means that during winter the colony will have an abundance of heater bees (non-foragers), and these homebody bees will be available to heat the next generation of bees, who, once again, will be raised in a warm enough temperature to become superior foragers—just in time for spring!

Forever Young

Honey bees usually start to forage around middle age. If more foragers are needed, young bees will step into the role, which works since the beehive's job system is remarkably flexible. Foraging is hard labor, though, and the stress of it shows. Once a bee progresses to foraging work, no matter how old she is, she begins to age quickly. After just two weeks her wings become worn, she starts to go bald, and she even begins to lose cognitive function. Yes, her brain function actually declines, and she loses her ability to learn new things.

Researchers wanted to know how the bees would adapt if the colony was put in the reverse position and more nurse bees were needed instead of foragers. To test this the researchers created an extreme situation: They removed all the young bees from the nest, leaving only the brood and the foragers. When the foragers came home and discovered this strange happening, they responded by adapting their workforce. Some foragers continued their work, while others returned to the housework of their youth, bringing the colony to a state of balance once again. Incredibly, the foragers who returned to nest duties not only reversed their job timeline but also reversed their aging! The researchers found that after just 10 days, 50 percent of them had significantly improved their ability to learn new things and had new proteins in their brains.

Plan B

When a colony loses its queen unexpectedly, the bees respond by making replacements. This is a period fraught with uncertainty. Despite the bees' efforts, the new queens may not survive. A queen could become sick in her cell or get eaten by a bird on her mating flight. If the worst should happen and their new queen also dies, the workers do have a plan B. Although the colony as they know it is doomed, the workers do their best to preserve its genes by laying their own eggs. Since worker bees cannot mate, each one of their eggs grows into a drone, capable of passing on his colony's genes through mating. The colony must use its last few weeks of life to make as many drones as possible, in hopes that one of them will successfully mate.

The bees seem to know that they may not succeed in raising a new queen, and they balance their risk by preparing for the worst. When a colony first loses its queen, they don't just build queen cells; they also construct drone comb. At the same time worker bees begin developing their ovaries. They do this even though they have active replacement queens growing in their hive. The bees don't stop their preparations for the worst until their new queen begins laying eggs.

Sentience

Some people believe that honey bees are like tiny robots whose preprogrammed brains only respond to stimuli via a set of hardwired rules or instincts. Others argue that bees have demonstrated mammal-like behaviors including problem-solving abilities and sophisticated emotions never thought possible in insects. The latter sentiment has led to a rash of recent research and books that explore the possibility that bees may be more like us than we realize. Are honey bees sentient?

Comb Cognition

While instincts certainly play a role in honey bee behavior, they may not govern every aspect of honey bee life. Many honey bee activities now appear to be at least partially learned (see page 101).

We have long assumed that comb construction is an innate skill that doesn't require any thought. This is understandable since the repeating pattern is so consistent and precise, it looks as if it's been manufactured by a machine. In fact some scientists were so reluctant to give honey bees credit for their incredible architecture that they hypothesized it was merely the result of tension. They suggested that the bees build not hexagonal cells but round cells, and that the hexagons only form as the wax cools and is pulled in different directions by the neighboring cells. This theory has since been debunked as subsequent research has revealed that the wax never gets hot enough for this effect to occur, and although the cell rims appear round during construction, the actual structure below the rims is hexagonal. But how much of comb building is instinct, and how much, if any, is the result of honey bee intelligence?

Although we don't yet understand how the bees maintain such consistent measurements across their comb, careful observation of comb building has found that it's a truly collaborative effort. Dozens of workers contribute to the construction of each cell. The bees work simultaneously on two rows of cells at a time. Each one spends an average of just 17 seconds shaping a cell before moving to an adjacent one. Builders use their mandibles to mechanically manipulate the wax as they continue the work of the bee that was there before them. If the previous bee makes a mistake, the one after corrects it. This suggests that the bees may be more like human typists, capable of reviewing and correcting their mistakes, rather than computer printers that move forward in their work regardless of any misprints.

Given the collaborative nature of comb building, some have wondered if older bees teach the younger ones how to do it. One study sought to answer this question by raising bee larvae in round cells. They found that even though the bees had never seen an example of hexagonal comb nor had contact with any bees who had, they still managed to build hexagons. The comb turned out a bit like it was created by a chatbot. The general idea of it was correct, but many of the details were off (for example, cell diameters were inconsistent). This suggests that while the use of the hexagon shape may be instinctual, the fine details of comb construction must be learned. Perhaps the comb-building strategy of sequential builders is a kind of apprenticeship that allows younger tradespeople to follow the example of older, more experienced ones.

Beekeepers who allow their bees to build natural comb may have noticed that their bees often build two or three separate combs on the same plane and then merge them together into one large one. The now- singular comb is often so seamless that you can't tell it was ever separate pieces. When studied in detail, however, the comb isn't always as perfect as perceived. Comb often has "imperfections," such as groups of non-hexagonal five- or seven-sided cells. These irregular cell

Two teams of comb builders work on adjacent combs on the same plane. Soon they will merge them into one continuous comb.

shapes are typically built when the bees are confronted with building challenges, such as the merging of two separate combs with unaligned cells. These findings again challenge the idea that honey bees are just comb-building automatons and instead suggests that they are capable of skilled architectural problem-solving.

Swiss entomologist Francis Huber explored this idea nearly two centuries ago when he noticed that his bees avoided attaching their comb to the glass in his observation hive. He conducted a series of experiments that tested the flexibility of comb building in the face of unusual challenges that would not be seen in nature. In a natural nest, the bees attach their comb to the uppermost surface and then gradually extend it downward. As each piece grows larger, they also secure it to the sides of the cavity for maximum stability. Huber first challenged his bees by placing them in a cavity with a glass ceiling. The bees adapted by building their comb from the bottom up. This is no small feat, because to build these unusual pyramid combs, the bees had to reverse their

Flawed Perfection

In 2022 one of my photos was published in a research paper. The paper explored cell shape abnormalities in comb, which I had unwittingly captured in a picture uploaded to my blog. To test the purpose of these irregular five- and seven-sided cells, the team of Canadian engineering researchers integrated the "defects" into a human-made beekeeping foundation and found that the abnormalities actually strengthened the overall structure of the comb. These irregular shapes appeared not only in the comb but also in the hexagonal lens pattern on the bees' eyes. My conclusion? Honey bees are perfect even when they are imperfect.

relationship with gravity. Imagine if a construction crew were forced to build a skyscraper from the top down!

Huber also tested what the bees would do if he gave them a cavity with a glass ceiling and a glass floor. They responded by building comb laterally between two walls, with no attachment at the top or bottom. As they extended their comb across the cavity toward their target wall on the opposite side, Huber added glass to that surface, too. Amazingly, the bees noticed the change in texture and altered their building trajectory in response. They began curving their comb so that it would attach to the wall running parallel to the comb instead. Not only does this experiment strengthen the idea that comb building is cognitive, but it also suggests that honey bees are like human engineers: They have the ability to plan ahead, perceive potential outcomes, and change their design to avoid undesirable results.

Honey Bee Emotions

Once considered a radical idea, the possibility that honey bees are capable of feeling emotion is something more and more researchers are exploring. As we learned on page 83, honey bees have impressive memories, which they use to optimize their foraging. When a forager finds a profitable group of flowers, she memorizes their color, pattern, and bloom time so that she can recognize this same flower type in the future. If she is ambushed by a crab spider during her visit, she remembers that, too, and may not return to those same flowers again.

Bees that escape predator attacks act a lot like human victims of trauma. In the same way a human might avoid dark alleys after being mugged in one, a traumatized bee avoids visiting the flower species she was on when she was attacked. If she does decide to visit that flower type again, her behavior is visibly altered. She approaches the flower more slowly and often hovers over it for longer than normal before landing. Sometimes she even appears to get spooked and quickly abandons the flower even though there is no threat. Could she be experiencing something like fear?

Most beekeepers agree: Bees have moods. One study sought to understand how bee behavior might change after bees survive a negative experience. The researchers trained a group of bees to expect a sugar reward from one smell and a bitter solution from another. When presented with the smell they associated with sugar, the bees extended their tongues, eager to drink. When presented with the bitter smell, they retracted their tongues, having learned that it would not taste good. In the second part of the study, half the smell-trained bees went through the bee-equivalent of a bad day: a simulated badger attack that involved shaking them vigorously for one full minute. Meanwhile, the other half of the group was given sugar treats.

Down but Not Out

A few years ago one of my beekeeping students called me for help. Her beehive had fallen over in a storm while she was out of town, and now, two weeks later, she wanted to know what could be done to salvage it. I arrived at the scene later that day and found that her four hive boxes had tumbled in such a way that, while they were still overlapping, they had landed completely upside down.

To my surprise, the bees had not abandoned their hive, though it was significantly altered. They were continuing to use their comb as normal and had even built queen cells with the intention to swarm. Because the queen cells had been built based on their new normal, when I returned the hive to its original position, they were upside down.

The researchers then presented the bees with the smells again, only this time they used a mixture of the two smells so that it was unclear to the bees whether they would receive a sweet reward or an unpleasant, bitter mouthful. The study found that the "bad day" bees were far less likely to extend their tongues in anticipation of sugar than the rewarded bees were. The bees displayed the same kind of pessimistic bias seen in humans, meaning that they were more likely to see the glass as half empty rather than half full if they were in a bad mood.

Transformation

Honey bees are transformative. They can change nectar into honey, beeswax into comb, and flowers into fruit. They can turn any cavity into a pristine palace or an architectural wonder. Under bees' influence, even something as unappealing as a trash bin can produce a golden treasure trove of honey. Honey bees even transform themselves, since each one begins as a larva and must go through metamorphosis to reach her adult form. Whether it's the flowers they pollinate or the cavities they inhabit, they seem to influence nearly everything they come into contact with, but perhaps none more than the humans who have come to love them.

Landscape Painters

Over time honey bees alter the landscape that surrounds them. By selectively pollinating their favorite flowers, they increase the number of seeds those plants produce. Not only does this help these bee-chosen plants to thrive, it gives them an advantage over their neighbors. Eventually these successful plants may outcompete other plants in the area and dominate the landscape. Of course this also benefits the honey bees. It's as if they restocked the neighborhood grocery store with only their favorite foods.

The honey bees' talent for reshaping their environment began with the very first bees to buzz on Earth. Evidence of the transformative landscaping abilities of ancient bees appeared in the fossil record as soon as they did, around 125 million years ago. These bees ushered in a new era of flowering plants and completely shifted what grows on Earth. During this time most plants reproduced by scattering their pollen to the wind in hopes that it would reach a neighboring plant, but the bees offered a more efficient solution. Instead of spending energy on mass pollen production, plants grew flowers and arranged to have the bees move the pollen for them. Flowers began to spring up everywhere and quickly developed bee-attracting features to distinguish themselves in a world of green.

Magnolias were among the first to unfold generous, white petals. The eye-catching petals made it easier for bees to spot them from the air and served as convenient landing pads. From there, flowers grew more complex and diverse, resulting in a dazzling array of colors and shapes. Some even provided a sweet nectar reward. With the help of bee pollination, flowering plants quickly replaced the dominant ferns and conifer trees of the time. Today they make up approximately 90 percent of the plants on Earth.

In response pollinators changed, too: They grew more hair, their tongues lengthened, and their diets changed to rely more on flowers. It's believed that the dietary change of a single wasp is what gave us the bee! When she stopped hunting prey to feed her young, opting to bring them pollen instead, she created a whole new branch of pollinators that would one day become 20 thousand different species of bee.

Even though a number of other animals contribute to pollination, when researchers map floral colors across natural landscapes around the world, they repeatedly find that their color distribution matches the color preferences of bees and not other pollinator groups such as birds.

This suggests that Earth's diverse bee species are truly what drove the evolution of floral shapes and colors. In this way they are nature's artists, painting our landscapes with their preferred palette.

Transformation of a Beekeeper

Honey bees changed me, too. When I first got into beekeeping, I had no particular interest in insects, but after a while, I started to wonder about the creatures adjacent to honey bees. I wanted to know more about other species of bees, other pollinators, and on and on. I began to slow down and pay attention to the tiny, unnoticed world that has been right alongside mine all along. I can now spend hours observing the insects in my garden, and I genuinely squeal with delight when I find a new type of visitor.

Honey bees have also helped me overcome my biases. Every time I want to dislike a less charismatic animal, like a fly or a wasp, I think of how some people dislike honey bees because they don't understand them. It makes me realize the potential of every creature, no matter how creepy or crawly. Most of all, honey bees connect me to my environment. They have made me aware of weather patterns, bloom times, and pesticides—I can see the effects of them in my own backyard. When nature is in balance my colonies thrive, and the whole garden seems to vibrate with a feeling that all is right with the world. Honey bees inspire me to make a better world for them and all the creatures we share this planet with.

Honey for Thought

It may be that our profound connection to honey bees began with the dawning of our own species. The first humans emerged around two million years ago and were equipped with brains significantly larger than those of their predecessors. The reason for this sudden increase in brain mass is still debated, but many agree that to sustain these new, metabolically expensive brains, early hominoids would have required an energy-rich food. Although much of the literature focuses on the dietary importance of meat and tubers, some anthropologists point out that honey and honey bee larvae are an ideal brain food.

Just imagine ancient humans breaking off a piece of comb and eating it on the go as if it were an energy bar. After all, honey is one of the most energy-dense foods available in nature, and with the added protein and fat from larvae, it's a highly nutritious snack. Honeycomb is such a prized part of the human diet that the ancient practice of honey hunting continues even today.

If our ancestors did learn to hunt for honey, not only might this explain what fueled our dramatic brain growth, but it may also have been what helped humans outcompete other species. A strong beehive would have acted like a magic well, providing a year-round food that simply regenerated itself after combs were taken. If the theory is true, then it's no wonder our bond with honey bees is so divine and enduring. Honey bees transformed our brains and made us human.

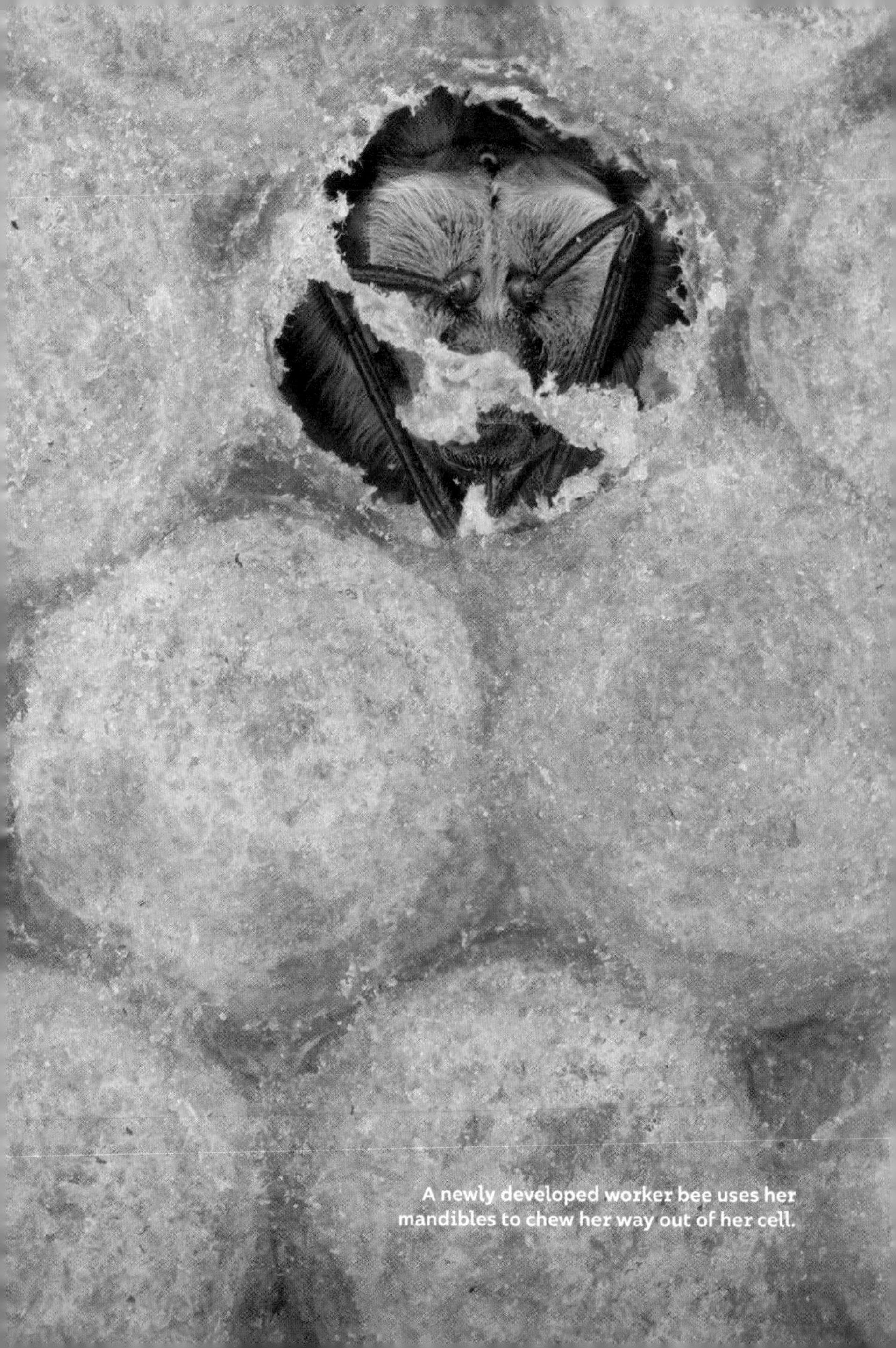

A newly developed worker bee uses her mandibles to chew her way out of her cell.

Bibliography by Chapter

Introduction

Hung, K. J., Kingston, J. M., Albrecht, M. et al. (2018). The worldwide importance of honey bees as pollinators in natural habitats. *Proceedings of the Royal Society B*, 285(1870). https://doi.org/10.1098/rspb.2017.2140

Chapter 2

Guard Bees

Downs, S. G., & Ratnieks, F. L. W. (2000). Adaptive shifts in honey bee (*Apis mellifera* L.) guarding behavior support predictions of the acceptance threshold model. *Behavioral Ecology, 11*(3), 326–333. https://doi.org/10.1093/beheco/11.3.326

Foraging: Day 18–Death

Klein, S., Pasquaretta, C., He, X.J. et al. (2019). Honey bees increase their foraging performance and frequency of pollen trips through experience. *Scientific Reports, 9*. https://doi.org/10.1038/s41598-019-42677-x

Chapter 3

Medicine

Gherman, B. I., Denner, A., Bobiş, O. et al. (2014). Pathogen-associated self-medication behavior in the honeybee *Apis mellifera*. *Behavioral Ecology and Sociobiology, 68*, 1777–1784. https://doi.org/10.1007/s00265-014-1786-8

Chapter 5

Learning

Wu, W., Moreno, A. M., Tangen, J. M., & Reinhard, J. (2013). Honeybees can discriminate between Monet and Picasso paintings. *Journal of Comparative Physiology A, 199*(1), 45–55. https://doi.org/10.1007/s00359-012-0767-5

Giurfa, M. (2021). Learning of sameness/difference relationships by honey bees: performance, strategies and ecological context. *Current Opinion in Behavioral Sciences, 37*, 1–6. https://doi.org/10.1016/j.cobeha.2020.05.008

Howard, S. R., Avarguès-Weber, A., Garcia, J. E. et al. (2018). Numerical ordering of zero in honey bees. *Science, 360*(6393), 1124–1126. https://doi.org/10.1126/science.aar4975

Memory

Grüter, C., Balbuena, M., & Farina, W. (2009). Retention of long-term memories in different age groups of honeybee (*Apis mellifera*) workers. *Insectes Sociaux, 56*, 385–387. https://doi.org/10.1007/s00040-009-0034-0

Touch

Smith, M. L., Koenig, P. A., & Peters, J. M. (2017). The cues of colony size: how honey bees sense that their colony is large enough to begin to invest in reproduction. *Journal of Experimental Biology, 220*(9), 1597–1605. https://doi.org/10.1242/jeb.150342

Forecasting the Weather

He, X. J., Tian, L. Q., Wu, X. B., & Zeng, Z. J. (2016). RFID monitoring indicates honeybees work harder before a rainy day. *Insect Science, 23*(1), 157–159. https://doi.org/10.1111/1744-7917.12298

Chapter 6

Dance Classes

Dong, S., Lin, T., Nieh, J. C., & Tan, K. (2023). Social signal learning of the waggle dance in honey bees. *Science, 379*(6636), 1015–1018. https://doi.org/10.1126/science.ade1702

The Stop Signal

Bell, H. C., Hsiung, K., Pasberg, P. et al. (2021). Responsiveness to inhibitory signals changes as a function of colony size in honeybees (*Apis mellifera*). *Journal of the Royal Society Interface, 18*(184). https://doi.org/10.1098/rsif.2021.0570

Grooming Invitation

Cini, A., Bordoni, A., Cappa, F. et al. (2020). Increased immunocompetence and network centrality of allogroomer workers suggest a link between individual and social immunity in honeybees. *Scientific Reports, 10*. https://doi.org/10.1038/s41598-020-65780-w

Chapter 7

Emergency Queens

Al-Kahtani, S. N., & Bienefeld, K. (2021). Strength surpasses relatedness—queen larva selection in honeybees. *PLoS ONE, 16*(8):e0255151. https://doi.org/10.1371/journal.pone.0255151

Chapter 8

Pesticides

DeGrandi-Hoffman, G., & Hagler, J. (2000). The flow of incoming nectar through a honey bee *(Apis mellifera* L.*)* colony as revealed by a protein marker. *Insectes Sociaux*, 47, 302–306. https://doi.org/10.1007/PL00001720

Pollution

Girling, R., Lusebrink, I., Farthing, E. et al. (2013). Diesel exhaust rapidly degrades floral odours used by honeybees. *Scientific Reports, 3*, 2779. https://doi.org/10.1038/srep02779

Parasites

Palmer-Young, E. C., Malfi, R., Zhou, Y. et al. (2023). Sunflower-associated reductions in varroa mite infestation of honey bee colonies. *Journal of Economic Entomology, 116*(1), 68–77. https://doi.org/10.1093/jee/toac196

Ardalani, H., Vidkjær, N. H., Kryger, P. et al. (2021). Metabolomics unveils the influence of dietary phytochemicals on residual pesticide concentrations in honey bees. *Environment International, 152*(106503). https://doi.org/10.1016/j.envint.2021.106503

Chapter 9

Energy Efficient Hives

Mitchell, D. (2019). Thermal efficiency extends distance and variety for honeybee foragers: analysis of the energetics of nectar collection and desiccation by *Apis mellifera*. *Journal of the Royal Society Interface, 16*(150). https://doi.org/10.1098/rsif.2018.0879

Hot and Cold Destiny

Tautz, J., Maier, S., Groh, C. et al. (2003). Behavioral performance in adult honey bees is influenced by the temperature experienced during their pupal development. *Proceedings of the National Academy of Sciences USA, 100*(12), 7343–7347. https://doi.org/10.1073/pnas.1232346100

Comb Cognition

Von Oelsen, G., & Rademacher, E. (1979). Studies on the building behavior of the honeybee (*Apis mellifica*). *Apidologie, 10*(2), 175–209. https://doi.org/10.1051/apido:19790208

Honey Bee Emotions

Bateson, M., Desire S., Gartside S. E., & Wright, G. A. (2011). Agitated honeybees exhibit pessimistic cognitive biases. *Current Biology, 21*(12),1070–1073. https://doi.org/10.1016/j.cub.2011.05.017

ACKNOWLEDGMENTS

To start with, I'd like to thank my former (now retired) editor, Deborah Burns, who pushed me to write this book even though I was in the haze of new motherhood. Next, I'd like to thank my current editor, Hannah Fries, for her support and patience despite all my missed deadlines. I'd also like to acknowledge the whole team at Storey Publishing for their excellent work and many contributions to the making of this book.

Many thanks also to all the researchers dedicated to furthering our understanding of honey bees—in particular, Dr. Heather Broccard-Bell for helping me work through some of the more complex studies and concepts I wished to explain in this book, and for always challenging my point of view with reasoned logic and maddening, unanswerable questions.

Most of all, I'd like to thank my friend, Mor Goldberger, without whom this book would not exist. Mor and I are old college friends who lost touch and reconnected over our shared struggles with new motherhood and postpartum depression. The challenge of writing a book during that period in my life was immense. Mor not only helped me come up with a strategy for finishing the book, she held me accountable. She texted me weekly, if not daily, asking for my word count progress. I truly could not have completed this book without her.

Lastly, I'd like to thank my husband, Tim O'Neil, for his unwavering faith in me and for all the cups of coffee he made me over the three-year writing period that finally resulted in a book.

Index

Page numbers in *italic* indicate illustrations or photographs.

Q

R

S

T

U

V

W

Z